GCSE Success

AQA

Combined Science

Foundation

Revision Guide

Complete Revision & Practice

Tom Adams
Dan Evans
Dan Foulder

AQA
required practicals -
you tube

Contents

Biology

Cell Biology

Required Practical 4

Transport Systems and Photosynthesis

Required Practical 2/3

Health, Disease and the Development of Medicines

Coordination and Control

Inheritance, Variation and Evolution

Ecosystems

Chemistry

Atomic Structure and the Periodic Table

Structure, Bonding and the Properties of Matter

Quantitative Chemistry

Physics

Energy

Waves

Electricity and Magnetism

Particle Model of Matter

Atomic Structure

Cells are the basis of life. All processes of life take place within them. There are many different kinds of specialised cell but they all have several common features. The two main categories of cells are **prokaryotic** (prokaryotes) and **eukaryotic** (eukaryotes).

Structure of cells

Prokaryotes

Prokaryotes are simple cells such as bacteria and archaebacteria. Archaebacteria are ancient bacteria with different types of cell wall and genetic codes to other bacteria. They include microorganisms that can use methane or sulfur as a means of nutrition (methanogens and thermoacidophiles).

Prokaryotic cells:

➤ are smaller than eukaryotic cells

➤ have cytoplasm, a cell membrane and a cell wall

➤ have genetic material in the form of a DNA loop, together with rings of DNA called plasmids

➤ do not have a nucleus.

The chart below shows the relative sizes of prokaryotic and eukaryotic cells, together with the magnification range of different viewing devices.

A bacterium

Plasmid DNA: a small, commonly circular, section of DNA that can replicate independently of chromosomal DNA

Ribosomes: sites of protein manufacture

Chromosomal DNA: the DNA of bacteria is not found within a nucleus and is usually found as one circular chromosome

Cytoplasm

Cell wall: provides structural support to the bacteria (is not made of cellulose)

Flagella: tail-like structures that rotate to help some bacteria move

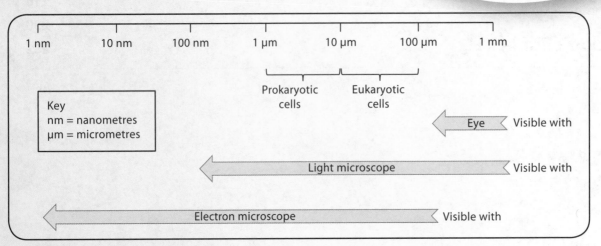

| 1 nm | 10 nm | 100 nm | 1 μm | 10 μm | 100 μm | 1 mm |

Prokaryotic cells

Eukaryotic cells

Key
nm = nanometres
μm = micrometres

Eye — Visible with

Light microscope — Visible with

Electron microscope — Visible with

Eukaryotes

Eukaryotic cells:

➤ are more complex than prokaryotic cells (they have a cell membrane, cytoplasm and genetic material enclosed in a nucleus)

➤ are found in animals, plants, fungi (e.g. toadstools, yeasts, moulds) and protists (e.g. amoeba)

➤ contain membrane-bound structures called **organelles**, where specific functions are carried out.

Here are the main organelles in plant and animal cells.

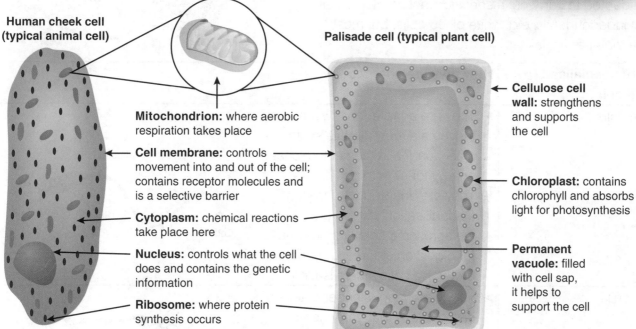

Human cheek cell (typical animal cell)

Palisade cell (typical plant cell)

Mitochondrion: where aerobic respiration takes place

Cell membrane: controls movement into and out of the cell; contains receptor molecules and is a selective barrier

Cytoplasm: chemical reactions take place here

Nucleus: controls what the cell does and contains the genetic information

Ribosome: where protein synthesis occurs

Cellulose cell wall: strengthens and supports the cell

Chloroplast: contains chlorophyll and absorbs light for photosynthesis

Permanent vacuole: filled with cell sap, it helps to support the cell

Plant cells tend to be more regular in shape than animal cells. They have additional structures: cell wall, sap vacuole and sometimes chloroplasts.

Create a poster illustrating the main features of bacterial, plant and animal cells (i.e. prokaryotes and eukaryotes).
➤ Use different colours for the main organelles, e.g. blue for the nucleus, orange for cytoplasm.
➤ Add labels that describe the functions of the organelles.

Keyword

Organelle ➤ A membrane-bound structure within a cell that carries out a particular function

1. List three examples of prokaryotes and three examples of eukaryotes.
2. How are prokaryotes and eukaryotes different in terms of genetic information?
3. All cells have a cell membrane. True or false?
4. Which organelle carries out the function of respiration?
5. Algae are a type of plant. Why do the cells contain chloroplasts?

Organisation and differentiation

Multicellular organisms need to have a coordinated system of structures so they can carry out vital processes, e.g. respiration, excretion, nutrition, etc.

Principles of organisation

Cells are the most basic unit of living organisms.

Tissues are collections of similar cells that work together to carry out the same function.

Organisation

An organ is formed when a group of tissues combine together and do a particular job.

Organs form organ systems, e.g. the intestines and the stomach are part of the digestive system.

Organ systems work together to make up a complex multicellular organism.

Cell specialisation

Animals and plants have many different types of cells. Each cell is adapted to carry out a specific function. Some cells can act independently, e.g. white blood cells, but most operate together as tissues.

Type of specialised animal cell	How they are adapted
Sperm cells	➤ They are adapted for swimming in the female reproductive system – mitochondria in the neck release energy for swimming. ➤ They are adapted for carrying out fertilisation with an egg cell – the acrosome contains enzymes for digestion of the ovum's outer protective cells at fertilisation.
Nerve cells	➤ They have long, slender extensions called axons that carry nerve impulses.
Muscle cells	➤ They are able to contract (shorten) to bring about the movement of limbs.

Type of specialised plant cell	How they are adapted
Root hair cells	➤ They have tiny, hair-like extensions. These increase the surface area of roots to help with the absorption of water and minerals.
Xylem	➤ They are long, thin, hollow cells. Their shape helps with the transport of water through the stem, roots and leaves.
Phloem	➤ They are long, thin cells with pores in the end walls. Their structure helps the cell sap move from one phloem cell to the next.

Cell differentiation and stem cells

Stem cells are found in animals and plants. They are unspecialised or **undifferentiated**, which means that they have the potential to become almost any kind of cell. Once the cell is fully specialised, it will possess sub-cellular organelles specific to the function of that cell. Embryonic stem cells, compared with those found in adults' bone marrow, are more flexible in terms of what they can become.

Animal cells are mainly restricted to repair and replacement in later life. Plants retain their ability to **differentiate** (specialise) throughout life.

Using stem cells in humans

Therapeutic cloning treats conditions such as diabetes. Embryonic stem cells (that can specialise into any type of cell) are produced with the same genes as the patient. If these are introduced into the body, they are not usually rejected.

Treating paralysis is possible using stem cells that are capable of differentiating into new nerve cells. The brain uses these new cells to transmit nervous impulses to the patient's muscles.

There are benefits and objections to using stem cells.

Benefits	Risks and objections
➤ Stem cells left over from in vitro fertilisation (IVF) treatment (that would otherwise be destroyed) can be used to treat serious conditions. ➤ Stem cells are useful in studying how cell division goes wrong, e.g. cancer. ➤ In the future, stem cells could be used to grow new organs for transplantation.	➤ Some people believe that an embryo at any age is a human being and so should not be used to grow cells or be experimented on. ➤ One potential risk of using stem cells is transferring viral infections. ➤ If stem cells are used in an operation, they might act as a reservoir of cancer cells that spread to other parts of the body.

Using stem cells in plants

Plants have regions of rapid cell division called **meristems**. These growth regions contain stem cells that can be used to produce clones cheaply and quickly.

Meristems can be used for:
➤ growing and preserving rare varieties to protect them from extinction
➤ producing large numbers of disease-resistant crop plants.

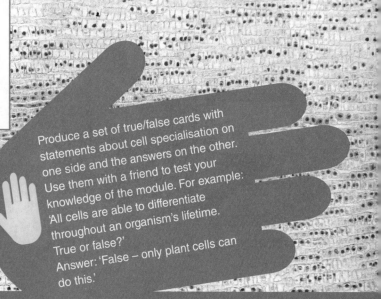

2

Produce a set of true/false cards with statements about cell specialisation on one side and the answers on the other. Use them with a friend to test your knowledge of the module. For example: 'All cells are able to differentiate throughout an organism's lifetime. True or false?' Answer: 'False – only plant cells can do this.'

1. In the body, what do you call an arrangement of different tissues that carry out a particular job?
2. Describe how a muscle cell is specialised to perform its function.
3. Which type of specialised plant cell is used to transport the sugars made in photosynthesis?
4. Give one benefit of and one objection to the scientific use of stem cells.

Microscopy and microorganisms

Microscopes

Microscopes:

➤ observe objects that are too small to see with the naked eye

➤ are useful for showing detail at cellular and sub-cellular level.

There are two main types of microscope: the **light microscope** and the **electron microscope**.

Light microscope

Eye piece

Objective lenses of different magnifications

Iris

Light source

Electron microscope

You will probably use a light microscope in your school laboratory.

The electron microscope was invented in 1931. It has increased our understanding of sub-cellular structures because it has much higher magnifications and resolution than a light microscope.

These are white blood cells, as seen through a transmission electron microscope. The black structures are nuclei.

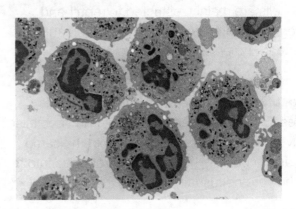

Here are some common units used in microscopy.

Measure	Scale	Symbol
1 metre		m
1 centimetre	$\frac{1}{100}$ th of a metre	cm
1 millimetre	$\frac{1}{1000}$ (a thousandth) of a metre	mm
1 micrometre	$\frac{1}{1\,000\,000}$ (a millionth) of a metre	μm
1 nanometre	$\frac{1}{1\,000\,000\,000}$ (a thousand millionth or a billionth) of a metre	nm

Comparing light and electron microscopes

Light microscope	Electron microscope
Uses light waves to produce images	Uses electrons to produce images
Low resolution	High resolution
Magnification up to ×1500	Magnification up to ×500 000 (2D) and ×100 000 (3D)
Able to observe cells and larger organelles	Able to observe small organelles
2D images only	2D and 3D images produced

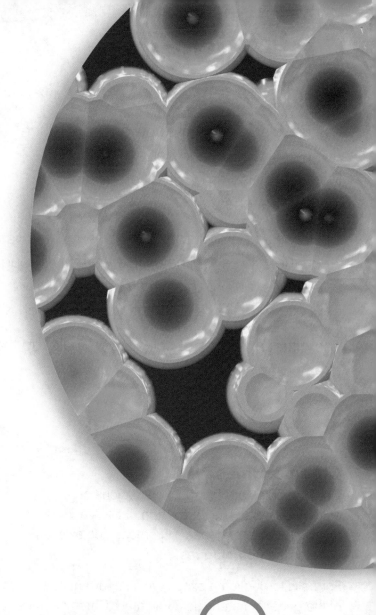

Magnification and resolution

Magnification measures how many times an object under a microscope has been made larger.

To calculate the magnifying power of a microscope, use this formula:

$$\text{magnification} = \frac{\text{size of image}}{\text{size of real object}}$$

You may be asked to carry out calculations using magnification.

> **Example:** A light microscope produces an image of a cell which has a diameter of 1500 μm. The cell's actual diameter is 50 μm. Calculate the magnifying power of the microscope.
>
> $$\text{magnification} = \frac{1500}{50} = \times 30$$

Resolution is the smallest distance between two points on a specimen that can still be told apart. The diagram below shows the limits of resolution for a light microscope. When looking through it you can see fine detail down to 200 nanometres. Electron microscopes can see detail down to 0.05 nanometres!

 200 nm

Staining techniques are used in light and electron microscopy to make organelles more visible.

➤ **Methylene blue** is used to stain the nuclei of animal cells for viewing under the light microscope.

➤ **Heavy metals** such as cadmium can be used to stain specimens for viewing under the electron microscope.

3

Design a poster illustrating the differences between the light and electron microscopes for a science lab notice board. Your poster should include the advantages of each type of microscope and diagrams to show the functions of each part.

➤ For example, you could include: 'Focusing wheel – adjusts the clarity of the image so it can be seen clearly.'

If you design your poster using a word processing or desktop publishing package, you can use specimen clipart to show the different types of view obtained.

1. List two differences between light and electron microscopes.
2. A certain sub-cellular structure is 150 nm across. Why can't it be seen using a light microscope?

Cell division

Multicellular organisms grow and reproduce using cell division and cell enlargement. Plants can also grow via differentiation into leaves, branches, etc.

Cell division begins from the moment of fertilisation when the zygote replicates itself exactly through **mitosis**. Later in life, an organism may use cell division to produce sex cells (**gametes**) in a different type of division called **meiosis**.

The different stages of cell division make up the **cell cycle** of an organism.

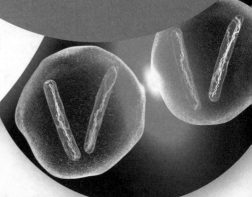

Chromosomes

Chromosomes are found in the nucleus of eukaryotic cells. They are made of DNA and carry a large number of genes. Chromosomes exist as pairs called **homologues**.

For cells to duplicate exactly, it is important that all of the genetic material is duplicated. Chromosomes take part in a sequence of events that ensures the genetic code is transmitted precisely and appears in the new **daughter cells**.

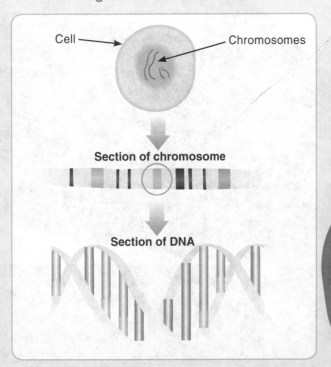

Cell → Chromosomes

Section of chromosome

Section of DNA

Fate of DNA during cell division

During the cell cycle, the genetic material (made of the **polymer** molecule, DNA) is doubled and then divided between the identical daughter cells.

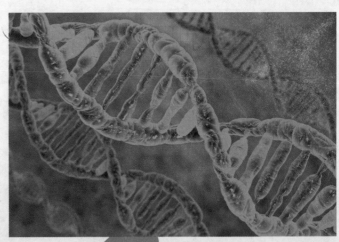

Use plasticine or playdough to create models showing mitosis and meiosis. Mitosis should show four stages and meiosis six stages.
You could:
➤ use different colours for the chromosomes and the cytoplasm
➤ try mixing the stages up and then re-arranging them from memory
➤ explain to a friend what is happening at each stage in terms of the chromosomes and cytoplasm.

Mitosis

Mitosis is where a **diploid** cell (one that has a complete set of chromosomes) divides to produce two more diploid cells that are genetically identical. Most cells in the body are diploid.

Humans have a diploid number of 46.

Mitosis produces new cells:
- for growth
- to replace old cells
- to repair damaged tissue
- for asexual reproduction.

Before the cell divides, the DNA is duplicated and other organelles replicate, e.g. mitochondria and ribosomes. This ensures that there is an exact copy of all the cell's content.

Mitosis – the cell copies itself to produce two genetically identical cells

Parent cell with two pairs of chromosomes.

Each chromosome replicates (copies) itself.

Chromosomes line up along the centre of the cell, separate into chromatids and move to opposite poles.

Each 'daughter' cell has the same number of chromosomes, and contains the same genes, as the parent cell.

Meiosis

Meiosis takes place in the testes and ovaries of sexually reproducing organisms and produces gametes (eggs or sperm). The gametes are called **haploid** cells because they contain half the number of chromosomes as a diploid cell. This chromosome number is restored during **fertilisation**.

Humans have a haploid number of 23.

Meiosis – the cell divides twice to produce four cells with genetically different sets of chromosomes

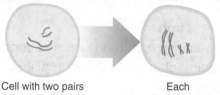

Cell with two pairs of chromosomes (diploid cell).

Each chromosome replicates itself.

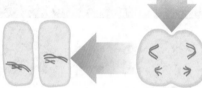

Chromosomes part company and move to opposite poles.

Cell divides for the first time.

Copies now separate and the second cell division takes place.

Four haploid cells (gametes), each with half the number of chromosomes of the parent cell.

Cancer

Cancer is a non-infectious disease caused by **mutations** in living cells.

Cancerous cells:
- divide in an uncontrolled way
- form **tumours**.

Benign tumours do not spread from the original site of cancer in the body. **Malignant tumour cells** invade neighbouring tissues. They spread to other parts of the body and form **secondary tumours**.

Making healthy lifestyle choices is one way to reduce the likelihood of cancer. These include:
- not smoking tobacco products (cigarettes, cigars, etc.)
- not drinking too much alcohol (causes cancer of the liver, gut and mouth)
- avoiding exposure to UV rays (e.g. sunbathing, tanning salons)
- eating a healthy diet (high fibre reduces the risk of bowel cancer) and doing moderate exercise to reduce the risk of obesity.

In addition to lifestyle choices, genetic risk factors can also influence the likelihood of some cancers.

1. Why do multicellular organisms carry out cell division?
2. Name the structures in the nucleus that carry genetic information.
3. What are the purposes of mitosis and meiosis?
4. Which structures are replicated during cell division?

Metabolism – respiration

Keywords

Exothermic reaction ➤ A reaction that gives out heat

Energy demand ➤ Energy required by tissues (particularly muscle) to carry out their functions

Oxygen debt ➤ The oxygen needed to remove lactic acid after exercise

Metabolism

Metabolism is the sum of all the chemical reactions that take place in the body.

The two types of metabolic reaction are:
➤ building reactions
➤ breaking-down reactions.

Building reactions

Building reactions require the input of energy. Examples include:
➤ converting glucose to starch in plants, or glucose to glycogen in animals
➤ the synthesis of lipid molecules

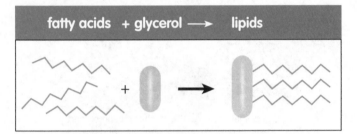

fatty acids + glycerol ⟶ lipids

➤ the formation of **amino acids** in plants (from glucose and nitrate ions) which, in turn, are built up into proteins.

Breaking down reactions

Breaking down reactions release energy. Examples include:
➤ breaking down amino acids to form **urea**, which is then excreted
➤ respiration.

Breaking down reactions produce waste energy in the form of heat (an **exothermic reaction**), which is transferred to the environment.

Respiration

Respiration continuously takes place in all organisms – the need to release energy is an essential life process. The reaction gives out energy and is therefore **exothermic**.

Aerobic respiration

Aerobic respiration takes place in cells. Oxygen and glucose molecules react and release energy. This energy is stored in a molecule called **ATP**.

glucose + oxygen ⟶ carbon dioxide + water

Energy is used in the body for many processes, including:
➤ muscle contraction (for movement)
➤ active transport
➤ transmitting nerve impulses
➤ synthesising new molecules
➤ maintaining a constant body temperature.

Anaerobic respiration

Anaerobic respiration takes place in the absence of oxygen and is common in muscle cells. It quickly releases a **smaller** amount of energy than aerobic respiration through the **incomplete breakdown** of glucose.

glucose ⟶ lactic acid + energy released

In plant and yeast cells, anaerobic respiration produces different products.

glucose ⟶ ethanol + carbon dioxide + energy released

This reaction is used extensively in the brewing and wine-making industries. It is also the initial process in the manufacture of spirits in a distillery.

Response to exercise

In animals, anaerobic respiration takes place when muscles are working so hard that the lungs and circulatory system cannot deliver enough oxygen to break down all the available glucose through aerobic respiration. In these circumstances the **energy demand** of the muscles is high.

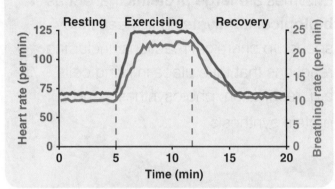

WS You may be asked to present data as graphs, tables, bar charts or histograms. For example, you could show data of breathing and heart rates on a line graph.

What does the line graph below tell you about breathing and pulse rates during recovery? Why is a line graph a good way to present the data?

Anaerobic respiration and recovery

Anaerobic respiration releases energy much faster over short periods of time. It is useful when short, intense bursts of energy are required, e.g. a 100 m sprint.

However, the incomplete oxidation of glucose causes **lactic acid** to build up. Lactic acid is toxic and can cause pain, cramp and a sensation of fatigue.

The lactic acid must be broken down quickly and removed to avoid cell damage and prolonged muscle fatigue.

➤ During exercise the body's heart rate, breathing rate and breath volume increase so that sufficient oxygen and glucose is supplied to the muscles, and so that lactic acid can be removed.

➤ This continues after exercise when deep breathing or panting occurs until all the lactic acid is removed. This repayment of oxygen is called **oxygen debt**.

1. Give one example of a building reaction and one of a breaking-down reaction.
2. Which type of respiration releases most energy – aerobic or anaerobic?
3. Give two uses of energy release in the body.
4. Name the products of anaerobic respiration in humans and in yeast.
5. Why do athletes pant after a race?

Metabolism – enzymes

Enzymes are **large proteins** that act as **biological** catalysts. This means they speed up chemical reactions, including reactions that take place in living cells, e.g. respiration, photosynthesis and protein synthesis.

Enzyme facts

Enzymes:

➤ are specific, i.e. one enzyme catalyses one reaction

➤ have an **active site**, which is formed by the precise folding of the enzyme molecule

➤ can be **denatured** by high temperatures and extreme changes in pH

➤ have an **optimum temperature** at which they work – for many enzymes this is approximately 37°C (body temperature)

➤ have an optimum pH at which they work – this varies with the site of enzyme activity, e.g. pepsin works in the stomach and has an optimum pH of 1.5 (acid), salivary amylase works best at pH 7.3 (alkaline).

Keywords

Catalyst ➤ A substance that controls the rate of a chemical reaction without being chemically changed itself

Active site ➤ The place on an enzyme molecule into which a substrate molecule fits

Denaturation ➤ When a protein molecule such as an enzyme changes shape and makes it unable to function

Substrate ➤ The molecule acted on by an enzyme

Kinetic energy ➤ Energy possessed by moving objects, e.g. reactant molecules such as enzyme and substrate molecules

Enzyme activity

Enzyme molecules work by colliding with **substrate** molecules and forcing them to break up or to join with others in synthesis reactions. The theory of how this works is called the **lock and key theory**.

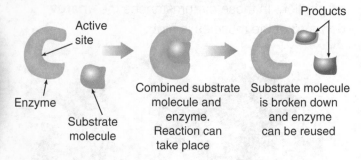

High temperatures denature enzymes because excessive heat vibrates the atoms in the protein molecule, putting a strain on the bonds and breaking them. This changes the shape of the active site.

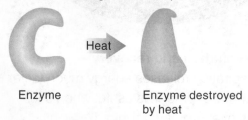

In a similar way, an extreme pH alters the active site's shape and prevents it from functioning.

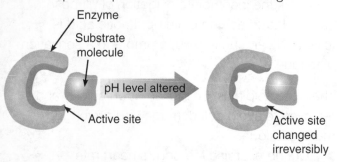

At lower than the optimal temperature, an enzyme still works but much more slowly. This is because the low **kinetic energy** of the substrate and enzyme molecules lowers the number of collisions that take place. When they do collide, the energy is not always sufficient to create a bond between them.

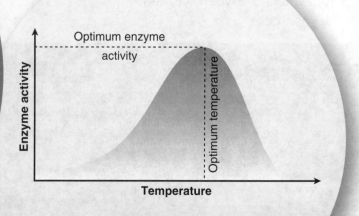

Enzymes in the digestive system

Enzymes in the digestive system help break down large nutrient molecules into smaller ones so they can be absorbed into the blood across the wall of the small intestine.

Enzyme types in the digestive system include carbohydrases, proteases and lipases.

Carbohydrases break down carbohydrates, e.g. **amylase**, which is produced in the mouth and small intestine.

> starch ⟶ maltose

Other carbohydrases break down complex sugars into smaller sugars.

Proteases break down protein, e.g. **pepsin**, which is produced in the stomach.

> protein ⟶ peptides ⟶ amino acids

Other enzymes in the small intestine complete protein breakdown with the production of amino acids.

Lipases, which are produced in the small intestine, break down lipids.

> lipid ⟶ fatty acids + glycerol

Bile

Bile is a digestive chemical. It is produced in the liver and stored in the gall bladder.
- ➤ Its alkaline pH neutralises hydrochloric acid that has been produced in the stomach.
- ➤ It **emulsifies** fat, breaking it into small droplets with a large surface area.
- ➤ Its action enables lipase to break down fat more efficiently.

What happens to digested food?

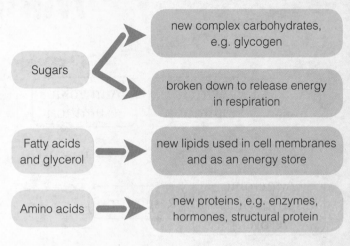

Sugars
- new complex carbohydrates, e.g. glycogen
- broken down to release energy in respiration

Fatty acids and glycerol → new lipids used in cell membranes and as an energy store

Amino acids → new proteins, e.g. enzymes, hormones, structural protein

WS During your course you will investigate how certain factors affect the rate of enzyme activity. These include temperature, pH and substrate concentration.

Design an investigation to discover how temperature affects the activity of amylase. Here are some guidelines.
- ➤ Iodine solution turns from a red-brown colour to blue-black in the presence of starch.
- ➤ You can measure amylase activity by timing how long it takes for iodine solution to stop turning blue-black.
- ➤ A water bath can be set up with a thermometer. Add cold or hot water to regulate the temperature.
- ➤ Identify the independent, dependent and control variables in the investigation. Write out your method in clear steps.

1. Name two factors that affect the rate of enzyme activity.
2. Describe, in simple terms, how lock and key theory explains enzyme action.
3. State two ways in which bile aids in the digestion of lipids.
4. Which molecules act as building blocks for protein polymers?

Mind map

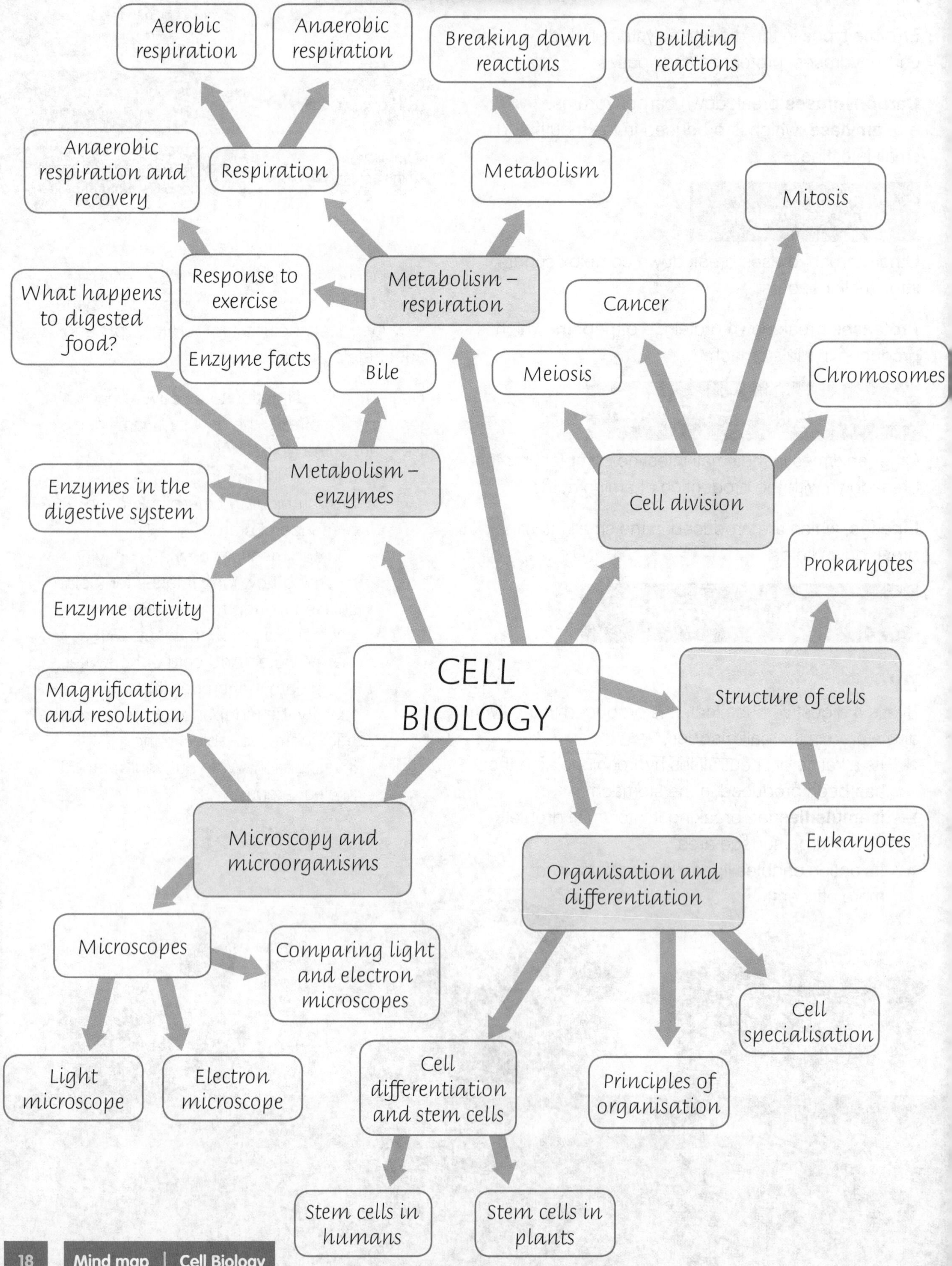

Aerobic respiration

Anaerobic respiration

Breaking down reactions

Building reactions

Anaerobic respiration and recovery

Respiration

Metabolism

Mitosis

Response to exercise

Metabolism – respiration

Cancer

What happens to digested food?

Enzyme facts

Bile

Meiosis

Chromosomes

Enzymes in the digestive system

Metabolism – enzymes

Cell division

Enzyme activity

Prokaryotes

Magnification and resolution

CELL BIOLOGY

Structure of cells

Eukaryotes

Microscopy and microorganisms

Organisation and differentiation

Microscopes

Comparing light and electron microscopes

Cell specialisation

Light microscope

Electron microscope

Cell differentiation and stem cells

Principles of organisation

Stem cells in humans

Stem cells in plants

Practice questions

1. Name the parts of the bacterial cell (**A–D**).   **(4 marks)**

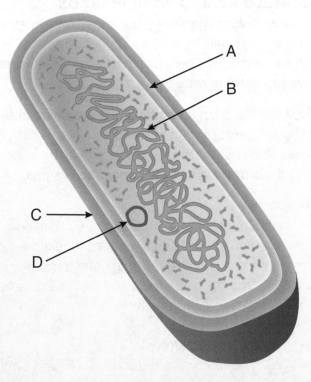

2. Lisa is playing football. She sprints the length of the pitch and scores a goal. However, she can barely celebrate because her legs have gone weak and she is panting heavily.

   a) Explain why Lisa's legs feel weak.   **(2 marks)**

   b) Why would Lisa be unable to play the entire game at such a fast pace?   **(1 mark)**

   c) Why does aerobic respiration yield more energy from glucose?   **(1 mark)**

3. In 1894, the lock and key theory of enzyme action was put forward by Emil Fischer. The diagram shows the first stage in the reaction between an enzyme and a reactant.

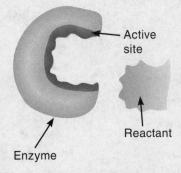

   a) What term describes the way that enzyme molecules change at high temperatures?   **(1 mark)**

   b) Explain how a drop in pH will affect the structure of the enzyme shown in the diagram.   **(2 marks)**

4. Red blood cells are one component of the human circulatory system. Explain why a human has these specialised cells but a paramecium (single-celled organism) does not.   **(2 marks)**

Cell transport

7

Diffusion

Living cells need to obtain oxygen, glucose, water, mineral ions and other dissolved substances from their surroundings. They also need to excrete waste products, such as carbon dioxide or urea. These substances pass through the cell membrane by **diffusion**.

Diffusion:

➤ is the (net) movement of particles in a liquid or gas from a region of high concentration to one of low concentration (down a **concentration gradient**)

➤ happens due to the random motion of particles past each other

➤ stops once the particles have completely spread out

➤ is passive, i.e. requires no input of energy

➤ can be increased in terms of rate by making the concentration gradient steeper, the diffusion path shorter, increasing the temperature or increasing the surface area over which the process occurs, e.g. having a folded cell membrane.

A protist called amoeba can absorb oxygen through diffusion.

Surface area to volume ratios

A **unicellular organism** (such as a protist) can absorb materials by diffusion directly from the environment. This is because it has a **large surface area to volume ratio**.

However, for a large, **multicellular organism**, the diffusion path between the environment and the inner cells of the body is long. Its large size also means that the **surface area to volume ratio** is **small**.

Adaptations

Multicellular organisms therefore need transport systems and specialised structures for exchanging materials, e.g. mammalian lungs and a small intestine, fish gills, and roots and leaves in plants. These increase diffusion efficiency in animals because they have:

➤ a large surface area

➤ a thin membrane to reduce the diffusion path

➤ an extensive blood supply for transport (animals)

➤ a ventilation system for gaseous exchange, e.g. breathing in animals.

In mammals, the individual air sacs in the lungs increase their surface area by a factor of thousands. Ventilation moves air in and out of the alveoli and the heart moves blood through the capillaries. This maintains the diffusion gradient. The capillary and alveolar linings are very thin, decreasing the diffusion path.

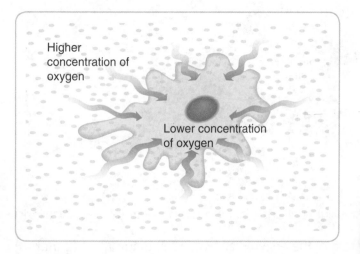

Higher concentration of oxygen

Lower concentration of oxygen

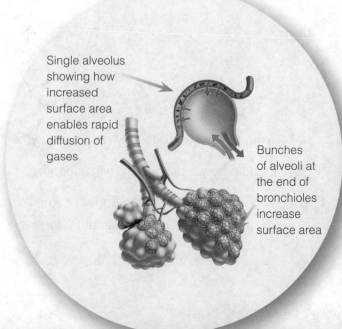

Single alveolus showing how increased surface area enables rapid diffusion of gases

Bunches of alveoli at the end of bronchioles increase surface area

Osmosis

Osmosis is a special case of diffusion that involves the movement of water only. There are two ways of describing osmosis.

1. The net movement of **water** from a region of **low** solute concentration to one of **high** concentration.
2. The movement down a **water potential gradient**.

Osmosis:

➤ occurs across a **partially permeable membrane**, so solute molecules cannot pass through (only water molecules can)
➤ occurs in all organisms
➤ is passive (in other words, requires no input of energy)
➤ allows water movement into root hair cells from the soil and between cells inside the plant
➤ can be demonstrated and measured in plant cells using a variety of tissues, e.g. potato chips, daffodil stems.

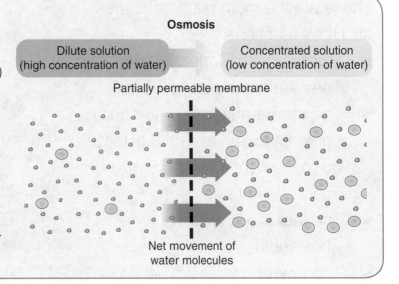

Osmosis

Dilute solution (high concentration of water) → Concentrated solution (low concentration of water)

Partially permeable membrane

Net movement of water molecules

Active transport

Substances are sometimes absorbed **against** a concentration gradient, i.e. from a low to a high concentration.

Active transport:

➤ requires the release of energy from respiration
➤ takes place in the small intestine in humans, where sugar is absorbed into the bloodstream
➤ allows plants to absorb mineral ions from the soil through root hair cells.

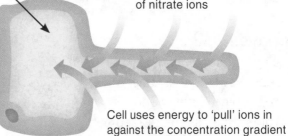

A cell absorbing ions by active transport

Root hair cell with high concentration of nitrate ions

Soil with lower concentration of nitrate ions

Cell uses energy to 'pull' ions in against the concentration gradient

WS During your course, you may investigate the effect of salt or sugar solutions on plant tissue.

Here is one experiment you could do.
1. Immerse raw potato cut into chips of equal length in sugar solutions of various concentration.
2. You will see the potato chips change in length depending on whether individual cells have lost or gained water.

Can you **predict** what would happen to the potato chips immersed in:

➤ concentrated sugar (e.g. 1 molar)
➤ medium concentration sugar (e.g. 0.5 molar)
➤ water (0 molar)?

Keywords

Surface area to volume ratio ➤ A number calculated by dividing the total surface area of an object by its volume. When the ratio is **high**, the efficiency of diffusion and other processes is **greater**

Water potential/diffusion gradient ➤ A higher concentration of particle numbers in one area than another; in living systems, these areas are often separated by a membrane or cell wall

Partially permeable membrane ➤ A membrane with microscopic holes that allows small particles through (e.g. water) but not large ones (e.g. sugar)

1. Which has the highest surface area to volume ratio – an elephant or a shrew?
2. Some plant tissue is placed in a highly concentrated salt solution. Explain why water leaves the cells.

A plant's system is made up of organs and tissues that enable it to be a **photosynthetic organism**.

These are the main plant structures and their functions.

➤ **Roots** absorb water and minerals. They anchor plants in the soil.

➤ The **stem** transports water and nutrients to leaves. It holds leaves up to the light for maximum absorption of energy.

➤ The **leaf** is the organ of photosynthesis.

➤ The **flower** makes sexual reproduction possible through pollination.

Plant tissues, organs and systems

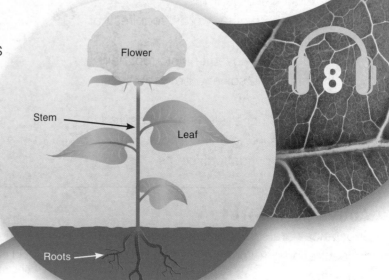

Flower

Stem

Leaf

Roots

8

Leaves

As this cross-section of a leaf shows, leaf tissues are adapted for efficient photosynthesis. The epidermis covers the upper and lower surfaces of the leaf and protects the plant against pathogens.

Upper epidermis – cells are thin and flat to allow light to pass through

Palisade layer (mesophyll) – contains many chloroplasts for light absorption. It is positioned near the top of the leaf to be nearer to sunlight

Mesophyll layer

Spongy layer (mesophyll) – air spaces allow efficient diffusion of gases

Guard cells – they open and close to control gas exchange

Lower epidermis

Stem and roots

Veins in the stem, roots and leaves contain tissues that transport water, carbohydrate and minerals around the plant.

➤ **Xylem tissue** transports water and mineral ions from the roots to the rest of the plant.

➤ **Phloem tissue** transports dissolved sugars from the leaves to the rest of the plant.

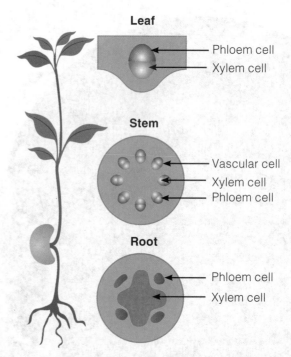

Leaf

Phloem cell
Xylem cell

Stem

Vascular cell
Xylem cell
Phloem cell

Root

Phloem cell
Xylem cell

Meristem tissue is found at the growing tips of shoots and roots.

Xylem, phloem and root hair cells

Xylem, phloem and root hair cells are adapted to their function.

Part of plant	Appearance	Function	How they are adapted to their function
Xylem	Hollow tubes made from dead plant cells (the hollow centre is called a lumen)	Transport water and mineral ions from the roots to the rest of the plant in a process called **transpiration**	The cellulose cell walls are thickened and strengthened with a waterproof substance called **lignin**
Phloem	Columns of living cells	**Translocate** (move) cell sap containing sugars (particularly sucrose) from the leaves to the rest of the plant, where it is either used or stored	Phloem have pores in the end walls so that the cell sap can move from one phloem cell to the next
Root hair cells	Long and thin; have hair-like extensions	Absorb minerals and water from the soil	Large surface area

Keywords

Photosynthetic organism ➤ Able to absorb light energy and manufacture carbohydrate from carbon dioxide and water

Transpiration ➤ Flow of water through the plant ending in evaporation from leaves

Lignin ➤ Strengthening, waterproof material found in walls of xylem cells

Translocation ➤ Process in which sugars move through the phloem

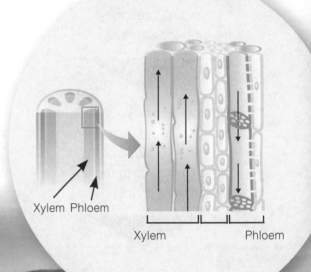

Xylem Phloem

Xylem   Phloem

If you snap a stick of celery open and pull the two halves apart, long 'strings' will pull away at the break. These are the long vascular tubes that carry water up to the celery's leaves.

➤ Cut a section from a stick of celery and look at it end-on. Can you see the vascular bundles that contain the xylem and phloem? To see the section more clearly, use a hand lens or magnifying glass.

➤ Try putting a celery stick (bottom end down) in a cup of food colouring overnight. When you cut the stem open the next day, you will see the coloured dye in the vascular bundles.

1. What part of the plant organ system allows water to enter the plant?
2. Why are there gaps between cells in the spongy mesophyll?
3. Why do xylem cells not have end walls?

Transport in plants

🎧 9

Transpiration

The movement of water through a plant, from roots to leaves, takes place as a transpiration stream. Once water is in the leaves, it diffuses out of the stomata into the surrounding air. This is called **(evapo)transpiration**.

> Water evaporates from the spongy mesophyll through the stomata.

> Water passes by osmosis from the xylem vessels in the leaf into the spongy mesophyll cells to replace what has been lost.

> This movement 'pulls' the column of water in that xylem vessel upwards.

> Water enters root hair cells by osmosis to replace water that has entered the xylem.

Factors affecting rate of transpiration

Evaporation of water from the leaf is affected by **temperature**, **humidity**, **air movement** and **light intensity**.

➤ **Increased temperature** increases the kinetic energy of molecules and removes water vapour more quickly.

➤ **Increased air movement** removes water vapour molecules.

➤ **Increased light intensity** increases the rate of photosynthesis. This in turn draws up more water from the transpiration stream, which maintains high concentration of water in the spongy mesophyll.

➤ **Decreasing atmospheric humidity** lowers water vapour concentration outside of the stoma and so maintains the concentration gradient.

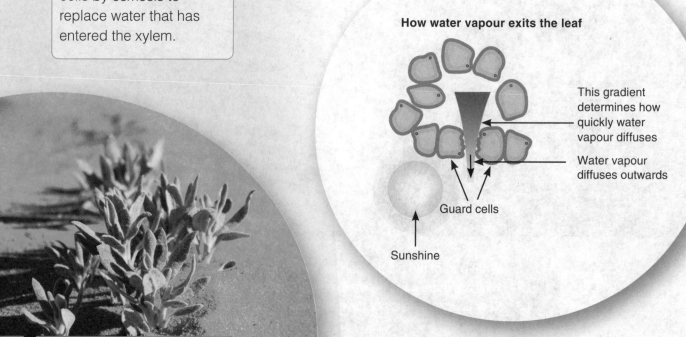

How water vapour exits the leaf

This gradient determines how quickly water vapour diffuses

Water vapour diffuses outwards

Guard cells

Sunshine

Opening and closing of stomata

Guard cells control the amount of water vapour that evaporates from the leaves and the amount of carbon dioxide that enters them.

➤ When light intensity is high and photosynthesis is taking place at a rapid rate, the sugar concentration rises in photosynthesising cells, e.g. palisade and guard cells.

➤ Guard cells respond to this by increasing the rate of water movement in the transpiration stream. This in turn provides more water for photosynthesis.

Stomata

 A **hypothesis** is an idea or explanation that you test through study and experiments. It should include a reason. For example: desert plants have fewer stomata than temperate plants **because** they need to minimise water loss.

➤ In an experiment investigating the factors that affect the rate of transpiration, a student plans to take measurements of weight loss or gain from a privet plant.

➤ Construct hypotheses for each of these factors: **temperature**, **humidity**, **air movement** and **light intensity**.

The first has been done for you: As temperature increases, the plant will lose mass/water more quickly **because** diffusion occurs more rapidly.

➤ Paint a thin layer of clear nail varnish onto the underside of a waxy leaf such as laurel – about 2 cm² should be sufficient.

➤ Allow the varnish to dry for at least 15 minutes. This will create a mould of the stomata as the liquid varnish fills the pores in the leaf.

➤ Peel the varnish strip off. Take care not to allow it to fold over or roll.

➤ Prepare it on a microscope slide with a drop of water and a cover slip.

➤ View it under medium power and you should see the stomata.

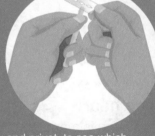

Compare different types of leaf, e.g. holly and privet, to see which leaves have most stomata. If you don't have a microscope at home, ask to use one at school.

Keywords

(Evapo)transpiration ➤ Evaporation of water from stomata in the leaf

1. What effect would **decreasing** air humidity have on transpiration?
2. In what circumstances might it be beneficial for plants to **close** their stomata?
3. How does water pass from xylem vessels into the leaf?
4. Describe how water passes from roots to leaves.

Transport in humans 1

🎧 10

Blood circulation

Blood moves around the body in a **double circulatory system**. In other words, blood moves twice through the heart for every full circuit. This ensures maximum efficiency for absorbing oxygen and delivering materials to all living cells.

The layout of the system

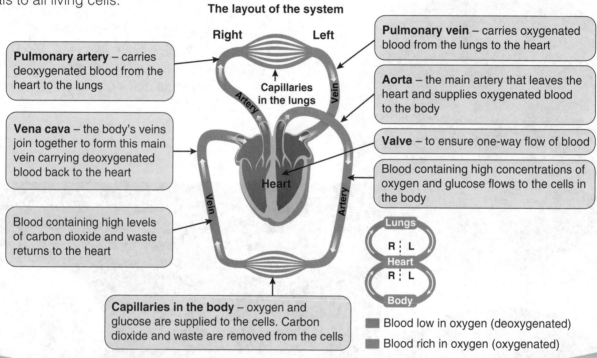

Right | Left

Pulmonary artery – carries deoxygenated blood from the heart to the lungs

Vena cava – the body's veins join together to form this main vein carrying deoxygenated blood back to the heart

Blood containing high levels of carbon dioxide and waste returns to the heart

Capillaries in the lungs

Artery | Vein

Heart

Vein | Artery

Capillaries in the body – oxygen and glucose are supplied to the cells. Carbon dioxide and waste are removed from the cells

Pulmonary vein – carries oxygenated blood from the lungs to the heart

Aorta – the main artery that leaves the heart and supplies oxygenated blood to the body

Valve – to ensure one-way flow of blood

Blood containing high concentrations of oxygen and glucose flows to the cells in the body

Lungs
R ┊ L
Heart
R ┊ L
Body

■ Blood low in oxygen (deoxygenated)
■ Blood rich in oxygen (oxygenated)

The heart

The heart is made of powerful muscles that contract and relax rhythmically in order to continuously pump blood around the body. The **heart muscle** is supplied with food (particularly glucose) and oxygen through the **coronary artery**.

The sequence of events that takes place when the heart beats is called the **cardiac cycle**.

1. The heart relaxes and blood enters both atria from the veins.
2. The atria contract together to push blood into the ventricles, opening the atrioventricular valves.
3. The ventricles contract from the bottom, pushing blood upwards into the arteries. The backflow of blood into the ventricles is prevented by the **semilunar valves**.

The left side of the heart is more muscular than the right because it has to pump blood further round the body. The right side only has to pump blood to the lungs and back.

Keywords

Coronary artery ➤ The blood vessel delivering blood to the heart muscle

Non-communicable ➤ Disease or condition that cannot be spread from person to person via pathogen transfer

Plaque ➤ Fatty deposits that can build up in arteries

Controlling the heartbeat

The heart is stimulated to beat rhythmically by pacemaker cells. The pacemaker cells produce impulses that spread across the atria to make them contract. Impulses are spread from here down to the ventricles, making them contract, pushing blood up and out.

Nerves connecting the heart to the brain can increase or decrease the pace of the pacemaker cells in order to regulate the heartbeat.

If a person has an irregular heartbeat, they can be fitted with an artificial, electrical pacemaker.

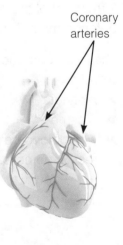

Pacemaker cells

Blood vessels

Blood is carried through the body in three types of vessel.
- **Arteries** have thick walls made of elastic fibres and muscle fibres to cope with the high pressure. The **lumen** (space inside) is small compared to the thickness of the walls. There are no valves.
- **Veins** have thinner walls. The lumen is much bigger compared to the thickness of the walls and there are valves to prevent the backflow of blood.
- **Capillaries** are narrow vessels with walls only one cell thick. These microscopic vessels connect arteries to veins, forming dense networks or **beds**. They are the only blood vessels that have permeable walls to allow the exchange of materials.

Artery

Lumen

Vein

Lumen

Valve

Capillary

Note: capillaries are much smaller than veins or arteries

Coronary heart disease

Coronary heart disease (CHD) is a **non-communicable** disease. It results from the build-up of **cholesterol**, leading to **plaques** laid down in the coronary arteries. This restricts blood flow and the artery may become blocked with a blood clot or **thrombosis**. The heart muscle is deprived of glucose and oxygen, which causes a **heart attack**.

The likelihood of plaque developing increases if you have a high fat diet. The risk of having a heart attack can be reduced by:
- eating a balanced diet and not being overweight
- not smoking tobacco
- lowering alcohol intake
- reducing salt levels in your diet
- reducing stress levels.

Coronary arteries

Healthy artery

Build-up of fatty material begins

Plaque forms

Plaque ruptures; blood clot forms

1. What is the function of valves in veins?
2. Describe how eating a diet high in fat can lead to a heart attack.

Transport in humans 2

Keywords

Haemoglobin ➤ Iron-containing molecule that binds to oxygen molecules in red blood cells

Ventilation ➤ Process of drawing air into and out of the lungs. It involves the ribs, intercostal muscles and diaphragm

Remedying heart disease

For patients who have heart disease, artificial implants called **stents** can be used to increase blood flow through the coronary artery.

Statins are a type of drug that can be taken to reduce blood cholesterol levels.

In some people, the heart valves may deteriorate, preventing them from opening properly. Alternatively, the valve may develop a leak.

This means that the supply of oxygenated blood to vital organs is reduced. The problem can be corrected by surgical replacement using a **biological** or **mechanical valve**.

When complete heart failure occurs, a heart transplant can be carried out. If a donor heart is unavailable, the patient may be kept alive by an artificial heart until one can be found. Mechanical hearts are also used to give the biological heart a rest while it recovers.

Blood as a tissue

Blood transports digested food and oxygen to cells and removes the cells' waste products. It also forms part of the body's defence mechanism.

The four components of blood are:
➤ platelets
➤ plasma
➤ white blood cells
➤ red blood cells.

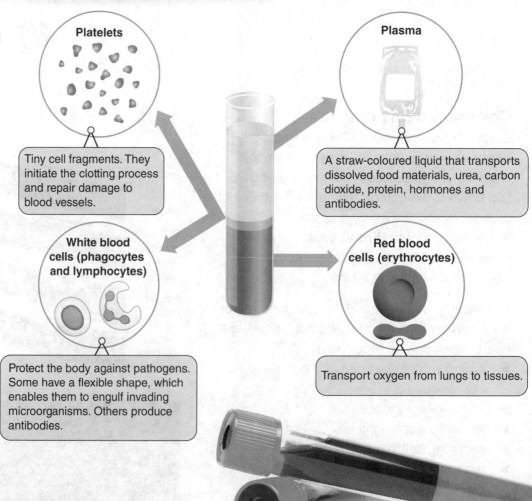

Platelets

Tiny cell fragments. They initiate the clotting process and repair damage to blood vessels.

Plasma

A straw-coloured liquid that transports dissolved food materials, urea, carbon dioxide, protein, hormones and antibodies.

White blood cells (phagocytes and lymphocytes)

Protect the body against pathogens. Some have a flexible shape, which enables them to engulf invading microorganisms. Others produce antibodies.

Red blood cells (erythrocytes)

Transport oxygen from lungs to tissues.

Oxygen transport

Red blood cells are small and have a biconcave shape. This gives them a large surface area to volume ratio for absorbing oxygen. When the cells reach the lungs, they absorb and bind to the oxygen in a molecule called **haemoglobin**.

> haemoglobin + oxygen ⇌ oxyhaemoglobin

Blood is then pumped around the body to the tissues, where the reverse of the reaction takes place. Oxygen diffuses out of the red blood cells and into the tissues.

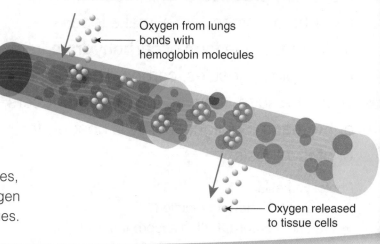

Transport of oxygen in red blood cells

Oxygen from lungs bonds with hemoglobin molecules

Oxygen released to tissue cells

The lungs

Humans, like many vertebrates, have lungs to act as a **gaseous exchange surface**.

Other structures in the **thorax** enable air to enter and leave the lungs (**ventilation**).

➤ The **trachea** is a flexible tube, surrounded by rings of cartilage to stop it collapsing. Air is breathed in via the mouth and passes through here on its way to the lungs.

➤ **Bronchi** are branches of the trachea.

➤ The **alveoli** are small air sacs that provide a large surface area for the exchange of gases.

➤ **Capillaries** form a dense network to absorb maximum oxygen and release carbon dioxide.

In the alveoli, **oxygen** diffuses down a concentration gradient. It moves across the thin layers of cells in the alveolar and capillary walls, and into the red blood cells.

For **carbon dioxide**, the gradient operates in reverse. The carbon dioxide passes from the blood to the alveoli, and from there it travels back up the air passages to the mouth.

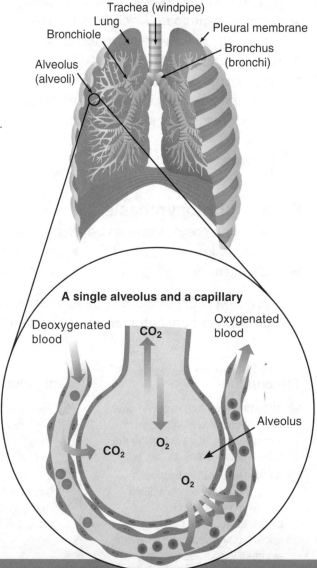

The lungs

Trachea (windpipe)
Lung
Bronchiole
Pleural membrane
Bronchus (bronchi)
Alveolus (alveoli)

A single alveolus and a capillary

Deoxygenated blood
CO_2
Oxygenated blood
Alveolus
CO_2
O_2
O_2

WS Scientists make observations, take measurements and gather data using a variety of instruments and techniques. Recording data is an important skill.

Create a table template that you could use to record data for the following experiment:

An investigation that involves measuring the resting and active pulse rates of 30 boys and 30 girls, together with their average breathing rates.

Make sure that:
➤ you have the correct number of columns and rows
➤ each variable is in a heading
➤ units are in the headings (so they don't need to be repeated in the body of the table).

1. What are the differences between lymphocytes and red blood cells?
2. Describe the route that would be travelled by a molecule of oxygen through the body until it reached a respiring muscle cell. State the cells, tissue, organs and processes that are involved.

Photosynthesis

Plants are **producers**. This means they can photosynthesise, i.e. make food molecules in the form of **carbohydrate** from the simple molecules, carbon dioxide and water. As such, they are the main producers of **biomass**. Sunlight energy is needed for photosynthesis.

Photosynthesis:

➤ is an **endothermic** reaction
➤ requires **chlorophyll** to absorb the sunlight; this is found in the **chloroplasts** of photosynthesising cells, e.g. palisade cells, guard cells and spongy mesophyll cells
➤ produces **glucose**, which is then respired for energy release or converted to other useful molecules for the plant
➤ produces **oxygen** that has built up in the atmosphere over millions of years; oxygen is vital for respiration in all organisms.

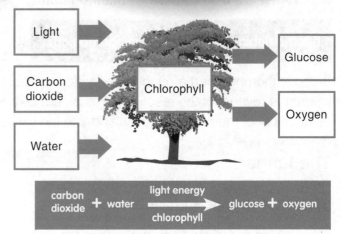

$$\text{carbon dioxide} + \text{water} \xrightarrow[\text{chlorophyll}]{\text{light energy}} \text{glucose} + \text{oxygen}$$

Rate of photosynthesis

The rate of photosynthesis can be affected by:

➤ temperature
➤ light intensity
➤ carbon dioxide concentration
➤ amount of chlorophyll.

In a given set of circumstances, **temperature**, **light intensity** and **carbon dioxide concentration** can act as **limiting factors**.

Temperature

1. As the temperature rises, so does the rate of photosynthesis. This means temperature is limiting the rate of photosynthesis.
2. As the temperature approaches 45°C, the enzymes controlling photosynthesis start to be denatured. The rate of photosynthesis decreases and eventually declines to zero.

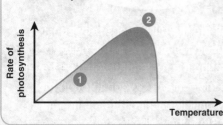

Light intensity

1. As the light intensity increases, so does the rate of photosynthesis. This means light intensity is limiting the rate of photosynthesis.
2. Eventually, the rise in light intensity has no effect on photosynthesis rate. Light intensity is no longer the limiting factor; carbon dioxide or temperature must be.

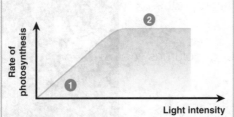

Carbon dioxide concentration

1. As carbon dioxide concentration increases, so does the rate of photosynthesis. Carbon dioxide concentration is the limiting factor.
2. Eventually, the rise in carbon dioxide concentration has no effect – it is no longer the limiting factor.

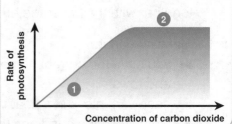

You need to understand that each factor has the potential to increase the rate of photosynthesis.

Biomass ➤ Mass of organisms calculated by multiplying their individual mass by the number that exist
Endothermic ➤ A change that requires the input of energy
Chlorophyll ➤ A molecule that gives plants their green colour and absorbs light energy
Limiting factor ➤ A variable that, if changed, will influence the rate of reaction most
Cellulose ➤ Large carbohydrate molecule found in all plants; an essential constituent of cell walls

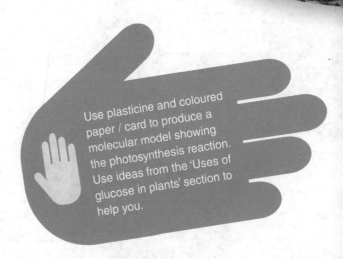

Uses of glucose in plants

The glucose produced from photosynthesis can be used immediately in respiration, but some is used to synthesise larger molecules: **starch**, **cellulose**, **protein** and **lipids**.

Starch is insoluble. So it is suitable for storage in leaves, stems or roots.

glucose individual sugar molecules	➡	starch huge, long chains of identical sugar molecules

Cellulose is needed for cell walls.

glucose individual sugar molecules	➡	cellulose long chains of sugar molecules; the chains are held together by weak bonds

Protein is used for the growth and repair of plant tissue, and also to synthesise enzyme molecules.

glucose individual sugar molecules	+	nitrates (from bacterial action in the soil)	➡	amino acids	➡	proteins huge, long chains of different amino acids

Lipids are needed in cell membranes, and for fat and oil storage in seeds.

glucose individual sugar molecules	➡	lipid structure

Use plasticine and coloured paper / card to produce a molecular model showing the photosynthesis reaction. Use ideas from the 'Uses of glucose in plants' section to help you.

1. Give two reasons why photosynthesis is seen as the opposite of respiration.
2. Describe how different concentrations of carbon dioxide can affect the rate of photosynthesis.

12

Mind map

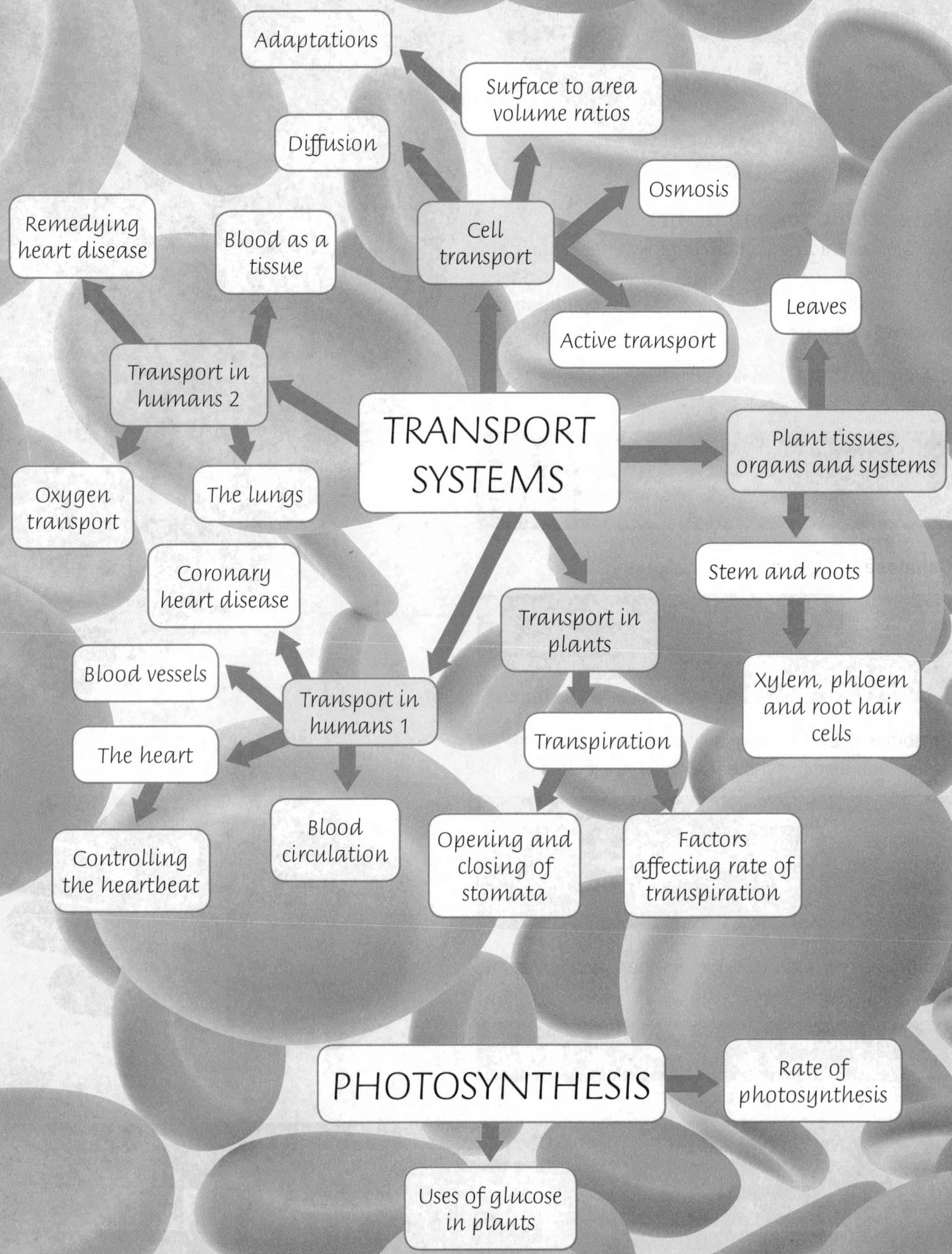

Adaptations

Surface to area volume ratios

Diffusion

Osmosis

Remedying heart disease

Blood as a tissue

Cell transport

Leaves

Active transport

Transport in humans 2

TRANSPORT SYSTEMS

Plant tissues, organs and systems

Oxygen transport

The lungs

Stem and roots

Coronary heart disease

Transport in plants

Xylem, phloem and root hair cells

Blood vessels

Transport in humans 1

The heart

Transpiration

Controlling the heartbeat

Blood circulation

Opening and closing of stomata

Factors affecting rate of transpiration

PHOTOSYNTHESIS

Rate of photosynthesis

Uses of glucose in plants

Practice questions

1. Which of the following are examples of osmosis? Tick (✓) the three correct options. **(3 marks)**

   a) Water evaporating from leaves ☐

   b) Water moving from plant cell to plant cell and back again ☐

   c) Mixing pure water and sugar solution ☐

   d) A pear submerged in a concentrated sugar solution losing water ☐

   e) Water moving from blood plasma to body cells ☐

   f) Sugar being absorbed from the intestine into the blood ☐

2. Emphysema is a lung disease that increases the thickness of the surface of the lungs for gas exchange and reduces the total area available for gas exchange.

   Two men did the same amount of exercise. One man was in good health and the other man had emphysema.

   The results are shown in the table.

	Healthy man	Man with emphysema
Total air flowing into lungs (dm³/min)	89.5	38.9
Oxygen entering blood (dm³/min)	2.5	1.2

   a) Which man had more oxygen entering his blood? **(1 mark)**

   b) Explain why the man with emphysema struggled to carry out exercise. **(2 marks)**

3. The diagram shows two types of blood vessel.

   A   B

   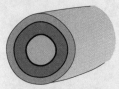 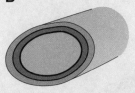

   a) Name each type of blood vessel. **(2 marks)**

   b) Explain why blood vessel A has a thick, elastic muscle wall. **(1 mark)**

   c) Why does blood vessel B have valves? **(1 mark)**

Non-communicable diseases

Keywords

Pathogen ➤ A microorganism that causes disease

Symptoms ➤ Physical or mental features that indicate a condition or disease, e.g. rash, high temperature, vomiting

Immune system ➤ A system of cells and other components that protect the body from pathogens and other foreign substances

Malnutrition ➤ A diet lacking in one or more food groups

13

Communicable diseases are caused by **pathogens** such as bacteria and viruses. They can be transmitted from organism to organism in a variety of ways. Examples include cholera and tuberculosis.

Non-communicable diseases are not primarily caused by pathogens. Examples are diseases caused by a poor diet, diabetes, heart disease and smoking-related diseases.

Health is the state of physical, social and mental well-being. Many factors can have an effect on health, including stress and life situations.

Risk factors

Non-communicable diseases often result from a combination of several **risk factors**.

➤ Risk factors produce an increased likelihood of developing that particular disease. They can be aspects of a person's lifestyle or substances found in the body or environment.

➤ Some of these factors are difficult to quantify or to establish as a definite **causal connection**.
So scientists have to describe their effects in terms of probability or likelihood.

The **symptoms** observed in the body may result from communicable and non-communicable components interacting.

A lowered **immune system** may make a person more vulnerable to infection.

Immune reactions caused by pathogens can trigger allergies such as asthma and skin rashes.

Symptoms

Viruses inhabiting living cells can change them into cancer cells.

Serious physical health problems can lead to **mental illness** such as depression.

Poor diet

People need a **balanced diet**.

If a diet does not include enough of the main food groups, **malnutrition** might result. Lack of correct vitamins leads to diseases such as **scurvy** and **rickets**. Lack of the mineral, iron, results in **anaemia**.

A high fat diet contributes to cardiovascular disease and high levels of salt increase blood pressure.

Alcohol

Drinking excess alcohol can impair brain function and lead to **cirrhosis** of the liver. It also contributes to some types of cancer and cardiovascular disease.

Smoking and drinking alcohol during pregnancy

Unborn babies receive nutrition from the mother via the placenta. Substances from tobacco, alcohol and other drugs can pass to the baby and cause **lower birth weight**, **foetal alcohol syndrome** and **addiction**.

Carcinogens

Exposure to **ionising radiation** (for example, X-rays, gamma rays) can cause cancerous tumours. Overexposure to UV light can cause skin cancer. Certain chemicals such as mercury can also increase the likelihood of cancer.

Smoking tobacco

Chemicals in tobacco smoke affect health.

- **Carbon monoxide** decreases the blood's oxygen-carrying capacity.
- **Nicotine** raises the heart rate and therefore blood pressure.
- **Tar** triggers cancer.
- **Particulates** cause **emphysema** and increase the likelihood of **lung infections**.

Weight/lack of exercise

Obesity and lack of exercise both increase the risk of developing **type 2 diabetes** and cardiovascular disease.

WS Interpreting complex data in graphs doesn't need to be difficult. This line graph shows data about smoking and lung cancer. Look for different patterns in it. For example:

- males have higher smoking rates in all years
- female cancer rates have increased overall since 1972.

Can you see any other patterns?

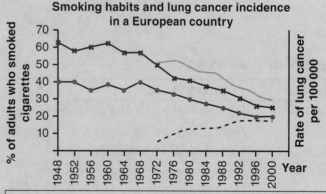

Smoking habits and lung cancer incidence in a European country

KEY
×××× Male smoking data ——— Male incidence of lung cancer
●●●● Female smoking data - - - - Female incidence of lung cancer

1. State three factors that cause cancer.
2. Give one consequence of a lowered immune system.

How do pathogens spread?

Pathogens are disease-causing microorganisms from groups of bacteria, viruses, fungi and protists. All animals and plants can be affected by pathogens. They spread in many ways, including:

- **droplet infection** (sneezing and coughing), e.g. flu
- **physical contact**, such as touching a contaminated object or person
- **transmission** by transferral of or contact **with bodily fluids**, e.g. hepatitis B
- **sexual transmission**, e.g. HIV, gonorrhoea
- **contamination of food or water**, e.g. Salmonella
- **animal bites**, e.g. rabies.

How do pathogens cause harm?

- Bacteria and viruses reproduce rapidly in the body.
- Viruses cause cell damage.
- Bacteria produce toxins that damage tissues.

These effects produce **symptoms** in the body.

How can the spread of disease be prevented?

The spread of disease can be prevented by:

- good hygiene, e.g. washing hands/whole body, using soaps and disinfectants
- destroying **vectors**, e.g. disrupting the life cycle of mosquitoes can combat malaria
- the isolation or quarantine of individuals
- vaccination.

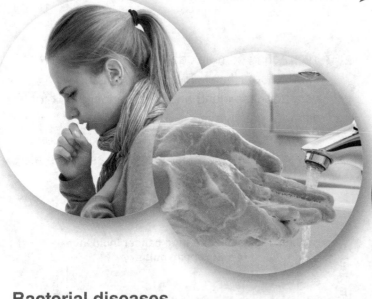

> Create cards showing information about the disease-causing microorganisms in this module. Each card should have the name of the disease at the top and then list the symptoms, mode of transmission, etc.

> Use the cards with a revision buddy. You could give a score for each category of information, e.g. contagion factor (how easily the disease is transmitted), severity of symptoms, etc.

Bacterial diseases

Disease	Transmission	Symptoms	Treatment/prevention
Salmonella	Contaminated food containing toxins from pathogens – these could be introduced from unhygienic food preparation techniques	Vomiting; fever; diarrhoea, stomach cramps	Anti-diarrhoeals and antibiotics; vaccinations for chickens
Gonorrhoea	Sexually transmitted	Thick yellow or green discharge from vagina or penis; pain on urination	Antibiotic injection followed by antibiotic tablets; penicillin is no longer effective against gonorrhoea; prevention through use of condoms

Viral and protist diseases

Disease	Transmission	Symptoms and notes	Treatment/prevention
Measles (viral)	Droplets from sneezes and coughs	Fever; red skin rash; fatal if complications arise	No specific treatment; vaccine is a highly effective preventative measure
HIV (viral)	Sexually transmitted; exchange of body fluids; sharing of needles during drug use	Flu-like symptoms initially; late-stage AIDS produces complications due to compromised immune system	Anti-retroviral drugs
Malaria (protist)	Via mosquito vector	Headache; sweats; chills and vomiting; symptoms disappear and reappear on a cyclical basis; further life-threatening complications may arise	Various anti-malarial drugs are available for both prevention and cure; prevention of mosquito breeding and use of mosquito nets

Malarial parasite

The **plasmodium** is a **protist** that causes malaria. It can reproduce asexually in the human host but sexually in the mosquito.

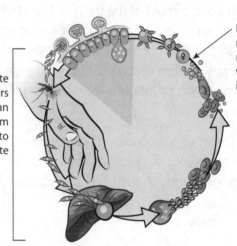

Parasite enters human from mosquito bite

Parasite re-enters mosquito when it feeds

Pathogens that affect plants

Disease	Pathogen	Appearance/effect on plants	Treatment/prevention
Rose black spot	Fungal disease – the fungal spores are spread by water and wind	Purple/black spots on leaves; these then turn yellow and drop early, leading to a lack of photosynthesis and poor growth 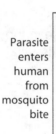	Apply a prevention fungicide and/or remove affected leaves Don't plant roses too close together. Avoid wetting leaves
Tobacco mosaic virus (TMV)	Widespread disease that affects many plants (including tomatoes)	'Mosaic' pattern of discolouration; can lead to lack of photosynthesis and poor growth	Remove infected plants Crop rotation Wash hands after treating plant

Keywords

Droplet infection ➤ Transmission of microorganisms through the aerosol (water droplets) produced through coughing and sneezing

Vectors ➤ Small organisms (such as mosquitoes or ticks) that pass on pathogens between people or places

1. Explain how reducing inspections of the food industry might affect the number of cases of Salmonella in a population.
2. How would treatment for HIV infection differ from that of Gonorrhoea?
3. Why is the tobacco mosaic virus so damaging to plants?

Human defences

Keywords

Epithelial ➤ A single layer of cells often found lining respiratory and digestive structures

Mucus ➤ Thick fluid produced in the lining of respiratory and digestive passages

Cilia ➤ Microscopic hairs found on the surface of epithelial cells; they 'waft' from side to side in a rhythmic manner

Antigen ➤ Molecular marker on a pathogen cell membrane that acts as a recognition point for antibodies

Antibodies ➤ Proteins produced by white blood cells (particularly lymphocytes). They lock on to antigens and neutralise them

Immunological memory ➤ The system of cells and cell products whose function is to prevent and destroy microbial infection

Non-specific defences

The body has a number of general or non-specific defences to stop pathogens multiplying inside it.

The skin covers most of the body – it is a **physical barrier** to pathogens. It also secretes antimicrobial peptides to kill microorganisms. If the skin is damaged, a clotting mechanism takes place in the blood preventing pathogens from entering the site of the wound.

Tears contain enzymes called **lysozymes**. Lysozymes break down pathogen cells that might otherwise gain entry to the body through tear ducts.

Hairs in the **nose** trap particles that may contain pathogens.

Tubes in the respiratory system (**trachea** and **bronchi**) are lined with special **epithelial** cells. These cells either produce a sticky, liquid **mucus** that traps microorganisms or have tiny hairs called **cilia** that move the mucus up to the mouth where it is swallowed.

The **stomach** produces **hydrochloric acid**, which kills microorganisms.

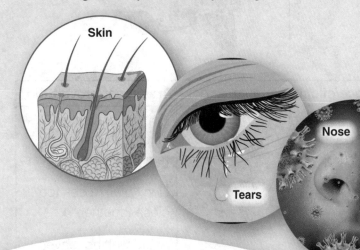

Skin

Tears

Nose

Epithelial cells

Cilia

Mucus produced here

Stomach

Phagocytes are a type of **white blood cell**. They move around in the bloodstream and body tissues searching for pathogens. When they find pathogens, they **engulf** and digest them in a process called **phagocytosis**.

White blood cell (phagocyte)

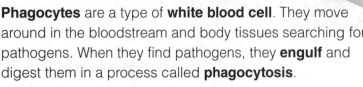

Microorganisms invade the body

The white blood cell surrounds and ingests the microorganisms

The white blood cell starts to digest the microorganisms

The microorganisms have been digested by the white blood cell

Specific defences

White blood cells called **lymphocytes** recognise molecular markers on pathogens called **antigens**. They produce **antibodies** that lock on to the antigens on the cell surface of the pathogen cell. The immobilised cells are clumped together and engulfed by phagocytes.

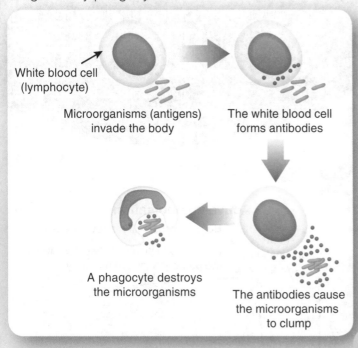

White blood cell (lymphocyte)

Microorganisms (antigens) invade the body

The white blood cell forms antibodies

A phagocyte destroys the microorganisms

The antibodies cause the microorganisms to clump

Some white blood cells produce **antitoxins** that neutralise the poisons produced from some pathogens.

Every pathogen has its own unique antigens. Lymphocytes make antibodies specifically for a particular antigen.

Example: Antibodies to fight TB will not fight cholera

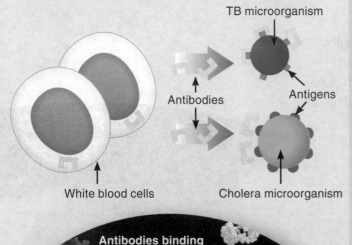

TB microorganism

Antibodies

Antigens

White blood cells

Cholera microorganism

Antibodies binding to virus

Active immunity

Once lymphocytes recognise a particular pathogen, the interaction is stored as part of the body's **immunological memory** through **memory lymphocytes**. These memory cells can produce the right antibodies much quicker if the same pathogen is detected again, therefore providing future protection against the disease. The process is called the **secondary response** and is part of the body's **active immunity**. Active immunity can also be achieved through vaccination.

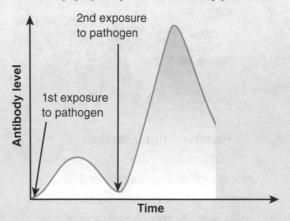

Memory lymphocytes and antibody production

2nd exposure to pathogen

Antibody level

1st exposure to pathogen

Time

WS Investigating the growth of pathogens in the laboratory involves culturing microorganisms. This presents hazards that require a **risk assessment**. A risk assessment involves taking into account the severity of each hazard and the likelihood that it will occur.

Any experiment of this type involves thinking about risks in advance. Here is an example of a risk assessment table for this investigation.

Hazard	Infection from pathogen	Scald from autoclave (a specialised pressure cooker for superheating its contents)
Risk	High	High
How to lower the risk	➤ Observe **aseptic technique**. ➤ Wash hands thoroughly before and after experiment. ➤ Store plates at a maximum temperature of 25°C.	➤ Ensure lid is tightly secured. ➤ Adjust heat to prevent too high a pressure. ➤ Wait for autoclave to cool down before removing lid.

1. Describe the process of phagocytosis.
2. Explain how active immunity can be developed in the body.

Fighting disease

Keywords

Sensitisation ➤ Cells in the immune system are able to 'recognise' antigens or foreign cells and respond by attacking them or producing antibodies

Herd immunity ➤ Vaccination of a significant portion of a population (or herd) makes it hard for a disease to spread because there are so few people left to infect. This gives protection to susceptible individuals such as the elderly or infants

Inhibition ➤ The effect of one agent against another in order to slow down or stop activity, e.g. chemical reactions can be slowed down using inhibitors. Some hormones are inhibitors

Vaccination

There are two types of vaccination.

Influenza Vaccine
10 ml

Passive immunisation

Antibodies are introduced into an individual's body, rather than the person producing them on their own. Some pathogens or toxins (e.g. snake venom) act very quickly and a person's immune system cannot produce antibodies quickly enough. So the person must be injected with the antibodies. However, this does not give long-term protection.

Active immunisation

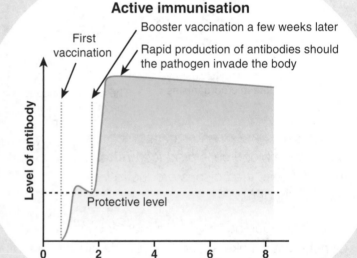

Booster vaccination a few weeks later

First vaccination

Rapid production of antibodies should the pathogen invade the body

Level of antibody

Protective level

Time (months)

Immunisation gives a person immunity to a disease without the pathogens multiplying in the body, or the person having symptoms.

1 A weakened or inactive strain of the pathogen is injected. The pathogen is heat-treated so it cannot multiply. The antigen molecules remain intact.

2 Even though they are harmless, the antigens on the pathogen trigger the white blood cells to produce specific antibodies.

3 As with natural immunity, **memory lymphocytes** remain **sensitised**. This means they can produce more antibodies very quickly if the same pathogen is detected again.

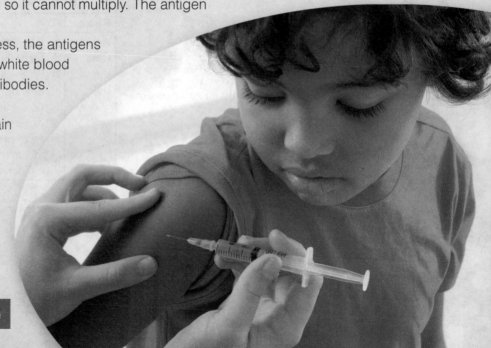

Antibiotics and painkillers

Diseases caused by bacteria (not viruses) can be treated using **antibiotics**, e.g. penicillin. Antibiotics are drugs that destroy the pathogen. Some bacteria need to be treated with antibiotics specific to them.

Antibiotics work because they **inhibit** cell processes in the bacteria but not the body of the host.

Viral diseases can be treated with **antiviral drugs**, e.g. swine flu can be treated with 'Tamiflu' tablets. It is a challenge to develop drugs that destroy viruses without harming body tissues.

Antibiotic resistance

Antibiotics are very effective at killing bacteria. However, some bacteria are **naturally resistant** to particular antibiotics. It is important for patients to follow instructions carefully and take the full course of antibiotics so that all the harmful bacteria are killed.

If doctors over-prescribe antibiotics, there is more chance of resistant bacteria surviving. These multiply and spread, making the antibiotic useless. **MRSA** is a bacterium that has become resistant to most antibiotics. These bacteria have been called 'superbugs'.

Painkillers

Painkillers or **analgesics** are given to patients to relieve symptoms of a disease, but they do not kill pathogens. Types of painkiller include paracetamol and ibuprofen. Morphine is another painkiller – it is a medicinal form of heroin used to treat extreme pain.

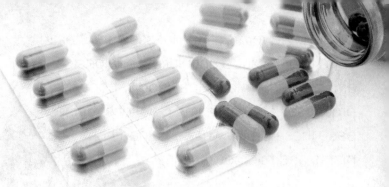

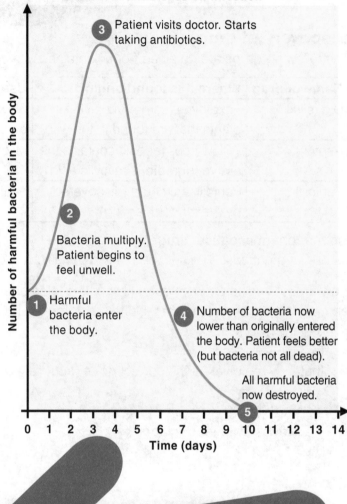

The effect of antibiotics on infection

Number of harmful bacteria in the body (y-axis)

Time (days) (x-axis): 0 1 2 3 4 5 6 7 8 9 10 11 12 13 14

3 – Patient visits doctor. Starts taking antibiotics.

2 – Bacteria multiply. Patient begins to feel unwell.

1 – Harmful bacteria enter the body.

4 – Number of bacteria now lower than originally entered the body. Patient feels better (but bacteria not all dead).

5 – All harmful bacteria now destroyed.

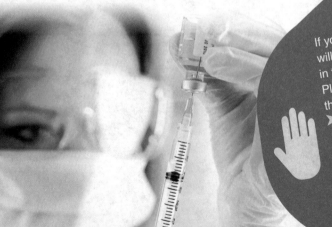

If you are revising with friends, try doing a role-play. This will help you visualise the actions of cells and processes in the immune system.
Plan and rehearse a scenario where a pathogen enters the human body and multiplies.
➤ Allocate roles for the pathogens and the various immune components: phagocytes, lymphocytes, antibodies, etc. You could even have a narrator.
➤ Mimic the actions of the various cells. How could you do this effectively?
➤ Ask someone to watch the role-play and give you feedback on what went well and what you could improve.

1. What is the difference between an antibiotic and an antibody?
2. What is the difference between an antiviral and an analgesic?
3. What can doctors and patients do to reduce the risk of antibiotic-resistant bacteria developing?

Drugs used for treating illnesses and health conditions include antibiotics, analgesics and other chemicals that modify body processes and chemical reactions. In the past, these drugs were obtained from plants and microorganisms.

Discovery and development of drugs

Discovery of drugs

The following drugs are obtained from plants and microorganisms.

Name of drug	Where it is found/origin	Use
Digitalis	Foxgloves (common garden plants that are found in the wild)	Slows down the heartbeat; can be used to treat heart conditions
Aspirin	Willow trees (aspirin contains the **active ingredient** salicylic acid)	Mild painkiller
Penicillin	Penicillium mould (discovered by Alexander Fleming)	Antibiotic

Modern **pharmaceutical drugs** are synthesised by chemists in laboratories, usually at great cost. The starting point might still be a chemical extracted from a plant.

New drugs have to be developed all the time to combat new and different diseases. This is a lengthy process, taking up to ten years. During this time the drugs are tested to determine:
➤ that they work
➤ that they are safe
➤ that they are given at the correct **dose** (early tests usually involve low doses).

New drugs made in laboratory

⬇

Drugs tested in laboratory for toxicity using cells, tissues and live animals

⬇

Clinical trials involving healthy volunteers and patients to check for side-effects

⬇

In addition to testing, **computer models** are used to predict how the drug will affect cells, based on knowledge about how the body works and the effects of similar drugs. There are many who believe this type of testing should be extended and that animal testing should be phased out.

Keywords

Active ingredient ➤ Chemical in a drug that has a therapeutic effect (other chemicals in the drug simply enhance flavour or act as bulking agents)
Pharmaceutical drug ➤ Chemicals that are developed artificially and taken by a patient to relieve symptoms of a disease or treat a condition
Placebo ➤ A substitute for the medication that does not contain the active ingredient

Clinical trials

Clinical trials are carried out on healthy volunteers and patients who have the disease. Some are given the new drug and others are given a **placebo**. The effects of the drug can then be compared to the effects of taking the placebo.

> **Blind trials** involve volunteers who do not know if they have been given the new drug or a placebo. This eliminates any psychological factors and helps to provide a fair comparison. (Blind trials are not normally used in modern clinical trials.)

> **Double blind trials** involve volunteers who are randomly allocated to groups. **Neither they nor the doctors/ scientists** know if they have been given the new drug or a placebo. This eliminates **all** bias from the test.

New drugs must also be tested against the best existing treatments.

 When studies involving new drugs are published, there is a **peer review**. This is where scientists with appropriate knowledge read the scientific study and examine the data to see if it is **valid**. Sometimes the trials are duplicated by others to see if similar results are obtained. This increases the **reliability** of the findings and filters out false or exaggerated claims.

Once a **consensus** is agreed, the paper is published. This allows others to hear about the work and to develop it further.

In the case of pharmaceutical drugs, clinical bodies have to decide if the drug can be **licensed** (allowed to be used) and whether it is **cost-effective**. This can be controversial because a potentially life-saving drug may not be used widely simply because it costs too much and/or would benefit too few people.

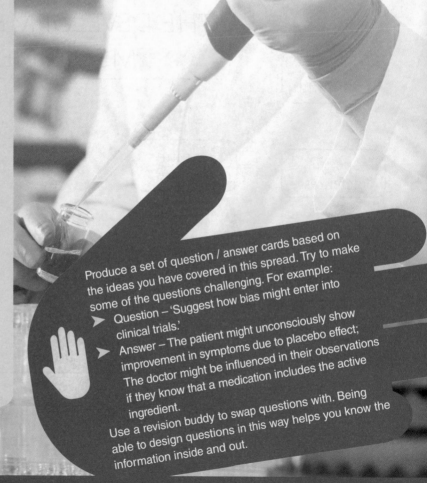

Produce a set of question / answer cards based on the ideas you have covered in this spread. Try to make some of the questions challenging. For example:

➤ Question – 'Suggest how bias might enter into clinical trials.'

➤ Answer – The patient might unconsciously show improvement in symptoms due to placebo effect; The doctor might be influenced in their observations if they know that a medication includes the active ingredient.

Use a revision buddy to swap questions with. Being able to design questions in this way helps you know the information inside and out.

1. What are the alternatives to testing drugs on animals?
2. Give two reasons why newly developed drugs need to be tested in clinical trials.

Mind map

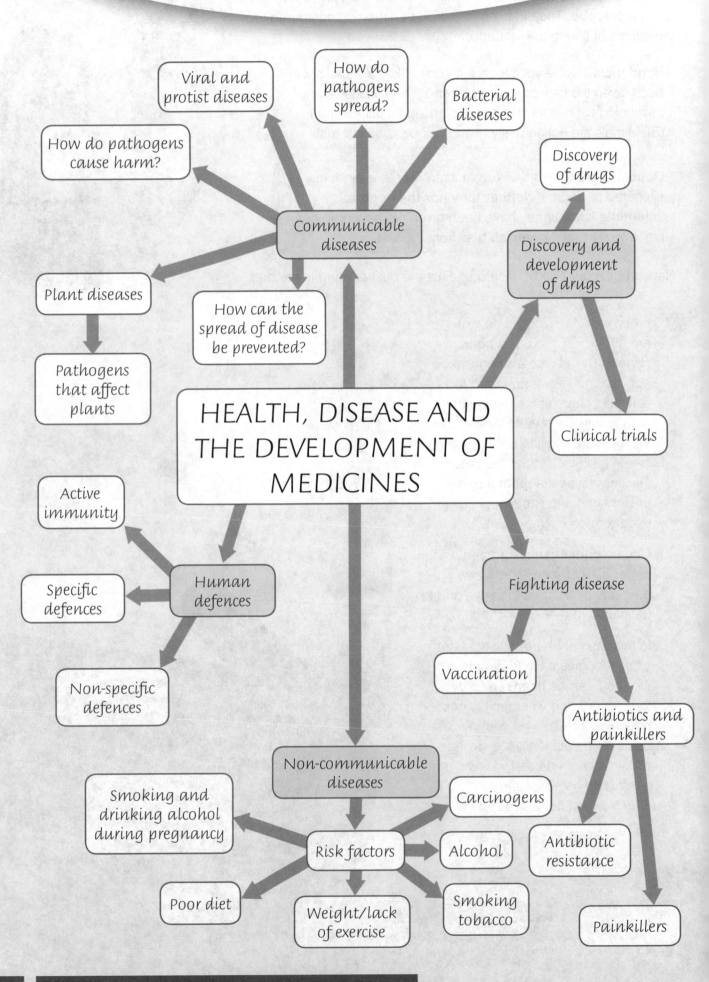

Viral and protist diseases

How do pathogens spread?

Bacterial diseases

How do pathogens cause harm?

Discovery of drugs

Communicable diseases

Discovery and development of drugs

Plant diseases

How can the spread of disease be prevented?

Pathogens that affect plants

Clinical trials

HEALTH, DISEASE AND THE DEVELOPMENT OF MEDICINES

Active immunity

Specific defences

Human defences

Fighting disease

Non-specific defences

Vaccination

Antibiotics and painkillers

Non-communicable diseases

Smoking and drinking alcohol during pregnancy

Carcinogens

Risk factors

Alcohol

Antibiotic resistance

Poor diet

Weight/lack of exercise

Smoking tobacco

Painkillers

Practice questions

1. Rani has caught flu and has been confined to bed for several days. Her mother is a health worker and was immunised against flu the previous month.

   a) Describe how the different types of blood component deal with the viruses in Rani's body.

      i)   phagocytes   **(1 mark)**

      ii) antibodies   **(1 mark)**

   b) Describe how the cells in Rani's mother's body responded to the vaccination she was given.

   In your answer, state what was in the vaccine and use the words **antigen**, **antibody** and **memory cells**.   **(4 marks)**

   c) After four days Rani is still unwell and she goes to the doctor. The doctor advises plenty of rest, regular intake of fluids and painkillers when necessary.

   Explain why the doctor doesn't prescribe antibiotics.   **(2 marks)**

   d) If Rani's symptoms continued, which other type of drug might she be prescribed?   **(1 mark)**

2. Two drugs called 'Redu' and 'DDD' have been developed to help obese people lose weight. Clinical trials are carried out on the two drugs. The results are shown in the table.

Drug	Number of volunteers in trial	Average weight loss in 6 weeks (kg)
Redu	3250	3.2
DDD	700	5.8
Placebo	2800	2.6

   a) The scientific team concluded that DDD was a more effective weight loss drug. Do you agree?

   Use the data in the table to give a reason for your answer.   **(2 marks)**

   b) The trial carried out was 'double blind'.

   Explain what this term means and why it is used.   **(2 marks)**

Homeostasis

Keywords

Endocrine system ➤ System of ductless organs that release hormones

Stimuli ➤ Changes in the internal or external environment that affect receptors

Homeostasis

The body has automatic control systems to maintain a constant internal environment (**homeostasis**). These systems make sure that cells function efficiently.

Homeostasis balances inputs and outputs to ensure that optimal levels of temperature, pH, water, oxygen and carbon dioxide are maintained. For example, even in the cold, homeostasis ensures that body temperature is regulated at about 37°C.

Control systems in the body may involve the nervous system, the **endocrine system**, or both. There are three components of control.

➤ **Effectors** cause responses that restore optimum levels, e.g. muscles and glands.

➤ **Coordination centres** receive and process information from the receptors, e.g. brain, spinal cord and pancreas.

➤ **Receptors** detect **stimuli** from the environment, e.g. taste buds, nasal receptors, the inner ear, touch receptors and receptors on retina cells.

Produce some cards with the following words on them: **Receptor, Control Centre** and **Effector**. Now, with a revision buddy, practice applying these ideas for particular examples.
➤ For example: **blood sugar**: receptor – cells in pancreas; control centre – pancreas; effector – insulin-producing cells in pancreas.
You could do the same for osmoregulation and body temperature.

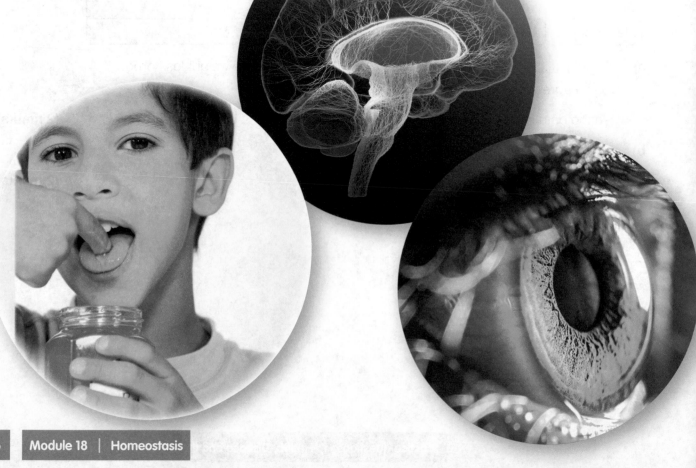

When taking measurements, the quality of the measuring instrument and a scientist's skill is very important to achieve **accuracy**, **precision** and **minimal error**.

Adrenaline levels in blood plasma are measured by a chromatography method called HPLC. This is often coupled to a detector that gives a digital readout.

This digital readout displays the concentration of adrenaline as 6.32. This means that the instrument is precise up to $\frac{1}{100}$ of a unit.

A less precise instrument might only measure down to $\frac{1}{10}$ of a unit, e.g. 6.3 (one decimal place).

A bar graph of some data generated from HPLC is shown below. The graph shows the **average concentration** of three different samples of blood. The average is taken from many individual measurements. The vertical error bars indicate the range of measurements (the difference between highest and lowest) obtained for each sample.

Sample C shows **the greatest precision** as the individual readings do not vary as much as the others. There is less error so we can be more confident that the average is closer to the **true value** and therefore more **accurate**.

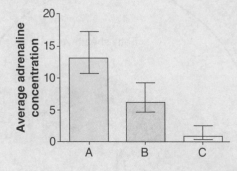

1. List the external stimuli that humans are sensitive to.
2. What is meant by 'homeostasis'?

The nervous system

Structure and function

The nervous system allows organisms to react to their surroundings and coordinate their behaviour.

The network of nerve cells that extends to all parts of the body are connected to the **central nervous system (CNS)**. This consists of the spinal cord and the **brain**.

The flow of **impulses** in the nervous system is carried out by nerve cells linking the receptor, coordinator (neurones and synapses in the CNS) and effector.

The main components of the nervous system

Brain

Spinal cord

Neurones outside of the CNS make up the rest of the nervous system.

CNS (brain and spinal cord)

Sense organ	Sensory neurone	Synapse	Relay neurone	Synapse	Motor neurone	Muscle
In the sense organ, **receptors** detect a change – either inside or outside the body. The change is a **stimulus**.	Conducts the impulse from the sense organ towards the CNS.	The gap between the sensory and relay neurones.	Passes the impulse on to a motor neurone.	The gap between the relay neurone and the motor neurone.	Passes the impulse on to the muscle (or gland).	The muscle responds by contracting, which results in a movement. Muscles and glands are examples of effectors.

Nerve cells or **neurones** are specially adapted to carry nerve impulses, which are electrical in nature. The impulse is carried in the long, thin part of the cell called the **axon**.

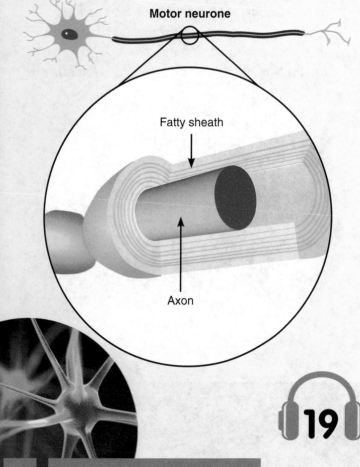

Motor neurone

Fatty sheath

Axon

There are three types of neurone.

Sensory neurones carry impulses from receptors to the CNS.

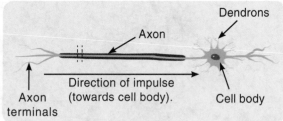

Dendrons

Axon

Direction of impulse (towards cell body).

Axon terminals

Cell body

Relay neurones make connections between neurones inside the CNS.

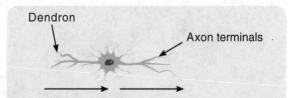

Dendron

Axon terminals

Impulse travels first towards, and then away from, cell body.

Motor neurones carry impulses from the CNS to muscles and glands.

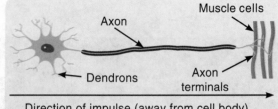

Muscle cells

Axon

Dendrons

Axon terminals

Direction of impulse (away from cell body).

19

Synapses

Synapses are junctions between neurones. They play an important part in regulating the way impulses are transmitted. Synapses can be found between different neurones, neurones and muscles, and between dendrites (the root-like outgrowths from the cell body).

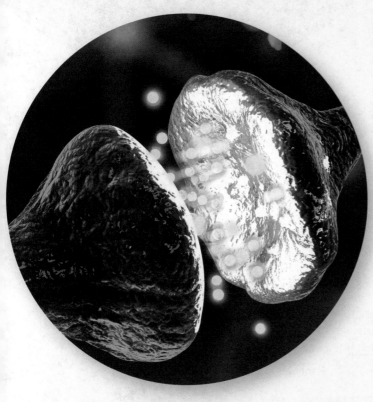

Keywords

Spinal cord ➤ Nervous tissue running down the centre of the vertebral column; millions of nerves branch out from it

Impulses ➤ Electrical signals sent down neurones that trigger responses in the nervous system

Reflex arcs

Reflex actions:
➤ are involuntary/automatic
➤ are very rapid
➤ protect the body from harm
➤ bypass conscious thought.

The pathway taken by impulses around the body is called a **reflex arc**. Examples include:
➤ opening and closing the pupil in the eye
➤ the knee-jerk response
➤ withdrawing your hand from a hot plate.

Here is the arc pathway for a pain response.

WS You may have to investigate the effect of factors on human reaction time.

For example, you could be asked to investigate a learned reflex by measuring how far up a ruler someone can catch it. The nearer to the zero the ruler is caught, the faster the reflex.

You could investigate factors such as:
➤ experience/practice at catching
➤ sound ➤ touch ➤ sight.

Can you design experiments to test these variables? Which factors will need to be kept the same?

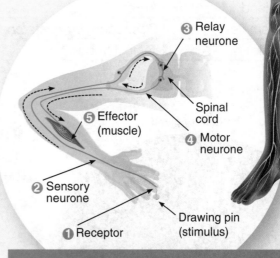

③ Relay neurone

⑤ Effector (muscle)

Spinal cord

④ Motor neurone

② Sensory neurone

Drawing pin (stimulus)

① Receptor

1. Name the long, thin extensions of nerve cells.
2. How do reflex arcs aid survival of an organism?
3. Design and draw a flow diagram to represent the pathway followed by impulses in the knee-jerk reflex. (The flow diagram at the beginning of this module should help you.)

The endocrine system

Keywords

Ductless gland ➤ A gland that does not secrete its chemicals through a tube. The pancreas is an exception to this rule as it contains a duct for delivering enzymes, but its hormones are released directly into the bloodstream

Glycogen ➤ Storage carbohydrate found in animals

Structure and function

The endocrine system is made up of glands that are **ductless** and secrete **hormones** directly into the bloodstream. The blood carries these chemical messengers to **target organs** around the body, where they cause an effect.

Hormones:

➤ are large, protein molecules

➤ interact with the nervous system to exert control over essential biological processes

➤ act over a longer time period than nervous responses but their effects are slower to establish.

Endocrine gland	Hormone(s) produced
Pituitary gland	TSH, ADH, FSH, LH, etc.
Pancreas	Insulin
Thyroid	Thyroxine
Adrenal gland	Adrenaline
Ovaries (female)	Oestrogen, progesterone
Testes (male)	Testosterone

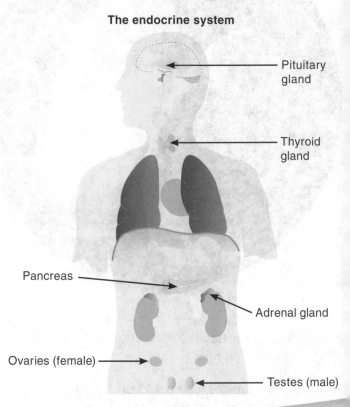

The endocrine system

- Pituitary gland
- Thyroid gland
- Pancreas
- Adrenal gland
- Ovaries (female)
- Testes (male)

Pituitary gland

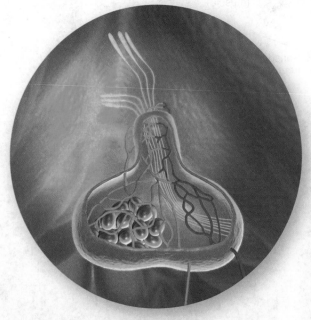

The pituitary gland

The pituitary is often referred to as the **master gland** because it secretes many hormones that control other processes in the body. Pituitary hormones often trigger other hormones to be released.

➤ Create a series of flashcards about controlling blood glucose levels. On each one, write a hormone, organ or effect relating to the control system.

➤ Shuffle the cards and put them face down.

➤ With a revision buddy, or on your own, pick up each card and explain how the particular component is involved in the control process.

Controlling blood glucose concentration

The control system for balancing blood glucose levels involves the **pancreas**.

The pancreas monitors the blood glucose concentration and releases hormones to restore the balance. When the concentration is too high, the pancreas produces insulin that causes glucose to be absorbed from the blood by all body cells, but particularly those in the liver and muscles. These organs convert glucose to **glycogen** for storage until required.

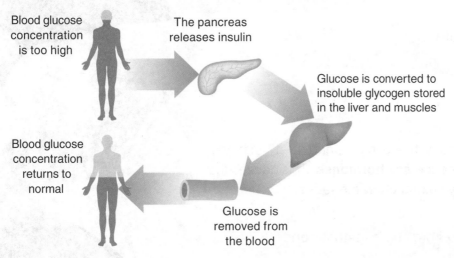

Blood glucose concentration is too high

The pancreas releases insulin

Glucose is converted to insoluble glycogen stored in the liver and muscles

Blood glucose concentration returns to normal

Glucose is removed from the blood

Diabetes

There are two types of diabetes.

Type I diabetes:

➤ is caused by the pancreas' inability to produce insulin
➤ results in dangerously high levels of blood glucose
➤ is controlled by delivery of insulin into the bloodstream via injection or a 'patch' worn on the skin
➤ is more likely to occur in people under 40
➤ is the most common type of diabetes in childhood
➤ is thought to be triggered by an auto-immune response where cells in the pancreas are destroyed.

Type II diabetes:

➤ is caused by fatty deposits preventing body cells from absorbing insulin; the pancreas tries to compensate by producing more and more insulin until it is unable to produce any more
➤ results in dangerously high levels of blood glucose
➤ is controlled by a low carbohydrate diet and exercise initially; it may require insulin in the later stages
➤ is more common in people over 40
➤ is a risk factor if you are obese.

1. Why is the pituitary called the master gland?
2. Where are the sex hormones of the body produced?
3. What effects does type II diabetes have on the body?

Hormones in human reproduction

Hormones play a vital role in regulating human reproduction, especially in the female **menstrual** cycle.

Puberty

During **puberty** (approximately 10–16 in girls and 12–17 in boys), the sex organs begin to produce **sex hormones**. This causes the development of **secondary sexual characteristics**.

In **males**, the primary sex hormone is **testosterone**.

During puberty, testosterone is produced from the testes and causes:
➤ production of sperm in testes
➤ development of muscles and penis
➤ deepening of the voice
➤ growth of pubic, facial and body hair.

In **females**, the primary sex hormone is **oestrogen**. Other sex hormones are **progesterone**, **FSH** and **LH**.

During puberty, oestrogen is produced in the ovaries and progesterone production starts when the menstrual cycle begins.

The secondary sexual characteristics are:
➤ ovulation and the menstrual cycle
➤ breast growth
➤ widening of hips
➤ growth of pubic and armpit hair.

To familiarise yourself with what the various hormones do in the menstrual cycle, you may find it helpful to do the following.
➤ On a computer, create a series of text boxes listing the organs, hormones and effects relating to the menstrual cycle. You could also include diagrams of the organs.
➤ Re-arrange the text boxes/diagrams so they are out of order.
➤ Test yourself or ask a revision buddy to put them back in the correct position/order. Alternatively, you could draw or write the organs, hormones and effects on paper and re-arrange them manually.

The menstrual cycle

A woman is fertile between the ages of approximately 13 and 50.

During this time, an egg is released from one of her ovaries each month and the lining of her uterus is replaced each month (approximately 28 days) to prepare for pregnancy.

Keywords

Menstruation ➤ Loss of blood and muscle tissue from the uterus wall during the monthly cycle
Follicle stimulating hormone ➤ A hormone produced by the pituitary gland that controls oestrogen production by the ovaries

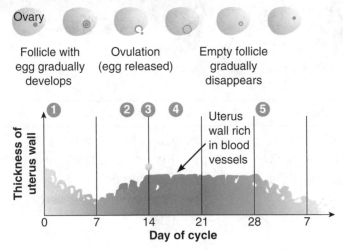

The menstrual cycle

Ovary

Follicle with egg gradually develops | Ovulation (egg released) | Empty follicle gradually disappears

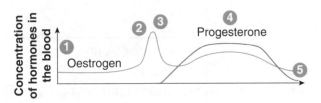

Uterus wall rich in blood vessels

Thickness of uterus wall

Day of cycle

0   7   14   21   28   7

Concentration of hormones in the blood

Oestrogen

Progesterone

1. Uterus lining breaks down (i.e. a period).
2. Repair of the uterus wall. Oestrogen causes the uterus lining to gradually thicken.
3. Egg released by the ovary.
4. Progesterone and oestrogen make the lining stay thick, waiting for a fertilised egg.
5. No fertilised egg so cycle restarts.

As well as oestrogen and progesterone, the two other hormones involved in the cycle are:
➤ **FSH** or **follicle stimulating hormone**, which causes maturation of an egg in the ovary
➤ **LH** or **luteinising hormone**, which stimulates release of an egg.

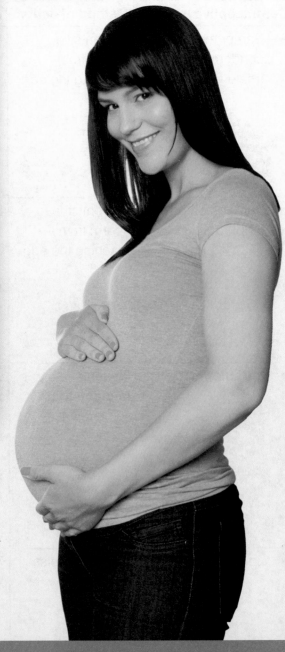

1. Name the four hormones involved in the menstrual cycle. What are their functions?
2. At approximately what stage in the menstrual cycle does the uterus lining repair itself?
3. Name one effect of testosterone in puberty.

Contraction

Fertility and the possibility of pregnancy can be controlled using non-hormonal and hormonal methods of **contraception**.

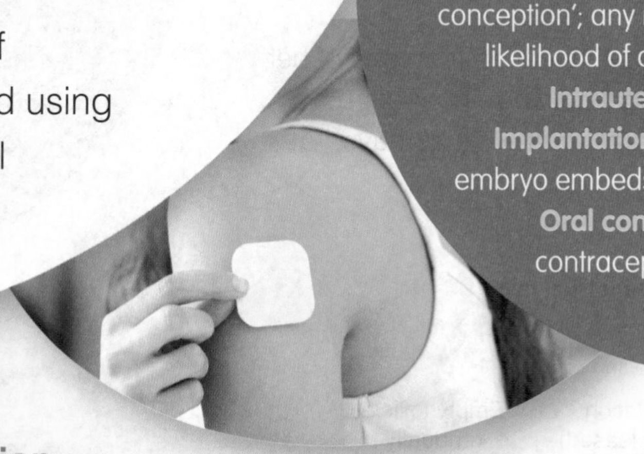

Keywords

Contraception ➤ Literally means 'against conception'; any method that reduces the likelihood of a sperm meeting an egg
Intrauterine ➤ Inside the uterus
Implantation ➤ Process in which an embryo embeds itself in the uterine wall
Oral contraceptive ➤ Hormonal contraceptive taken in tablet form

Non-hormonal contraception

Contraceptive method	Method of action	Advantages	Disadvantages
➤ Barrier method – condom (male + female) 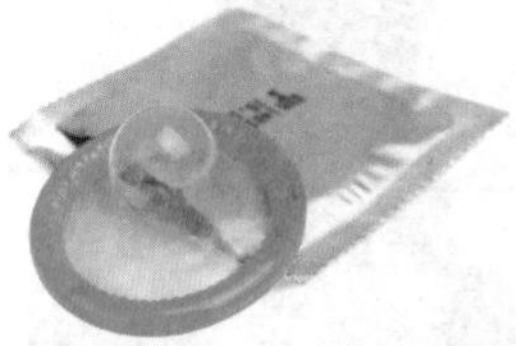	Prevents the sperm from reaching the egg	82% effective ➤ Most effective against STIs	➤ Can only be used once ➤ May interrupt sexual activity ➤ Can break ➤ Women may be allergic to latex
➤ Barrier method – diaphragm	Prevents the sperm from reaching the egg	88% effective ➤ Can be put in place right before intercourse or 2–3 hours before ➤ Don't need to take out between acts of sexual intercourse	➤ Increases urinary tract infections ➤ Doesn't protect against STIs
➤ **Intrauterine** device 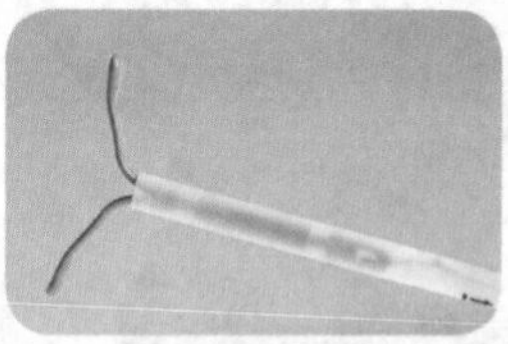	Prevents **implantation** – some release hormones	99% effective ➤ Very effective against pregnancy ➤ Doesn't need daily attention ➤ Comfortable ➤ Can be removed at any time	➤ Doesn't protect against STIs ➤ Needs to be inserted by a medical practitioner ➤ Higher risk of infection when first inserted ➤ Can have side effects such as menstrual cramping ➤ Can fall out and puncture the uterus (rare)
➤ Spermicidal agent	Kills or disables sperm	72% effective ➤ Cheap	➤ Doesn't protect against STIs ➤ Needs to be reapplied after one hour ➤ Increases urinary tract infections ➤ Some people are allergic to spermicidal agents

➤ Abstinence ➤ Calendar method 	Refraining from sexual intercourse when an egg is likely to be in the oviduct	76% effective ➤ Natural ➤ Approved by many religions ➤ Woman gets to know her body and menstrual cycles	➤ Doesn't protect against STIs ➤ Calculating the ovulation period each month requires careful monitoring and instruction ➤ Can't have sexual intercourse for at least a week each month
➤ Surgical method	Vasectomy and female sterilisation	99% effective ➤ Very effective against pregnancy ➤ One-time decision providing permanent protection	➤ No protection against STIs ➤ Need to have minor surgery ➤ Permanent

Hormonal contraception

Contraceptive method	Method of action	Advantages	Disadvantages
➤ **Oral contraceptive**	Contains hormones that inhibit FSH production, so eggs fail to mature	91% effectiveness ➤ Very effective against pregnancy if used correctly ➤ Makes menstrual periods lighter and more regular ➤ Lowers risk of ovarian and uterine cancer, and other conditions ➤ Doesn't interrupt sexual activity	➤ Doesn't protect against STIs ➤ Need to remember to take it every day at the same time ➤ Can't be used by women with certain medical problems or by women taking certain medications ➤ Can occasionally cause side effects
➤ Hormone injection ➤ Skin patch ➤ Implant	Provides slow release of progesterone; this inhibits maturation and release of eggs	91–99% effectiveness depending on method used ➤ Lasts over many months or years ➤ Light or no menstrual periods ➤ Doesn't interrupt sexual activity	➤ Doesn't protect against STIs ➤ May require minor surgery (for implant) ➤ Can cause side effects

The percentage figures in the contraception tables are based on users in a whole population, regardless of whether they use the method correctly. If consistently used correctly, the percentage effectiveness of each method is usually higher. Some methods, such as the calendar method, are more prone to error than others.

1. Which contraceptives might not be suitable for a woman suffering from high blood pressure?
2. Why are condoms effective against the spread of HIV?
3. List two advantages of the hormone skin patch as a contraceptive.

Mind map

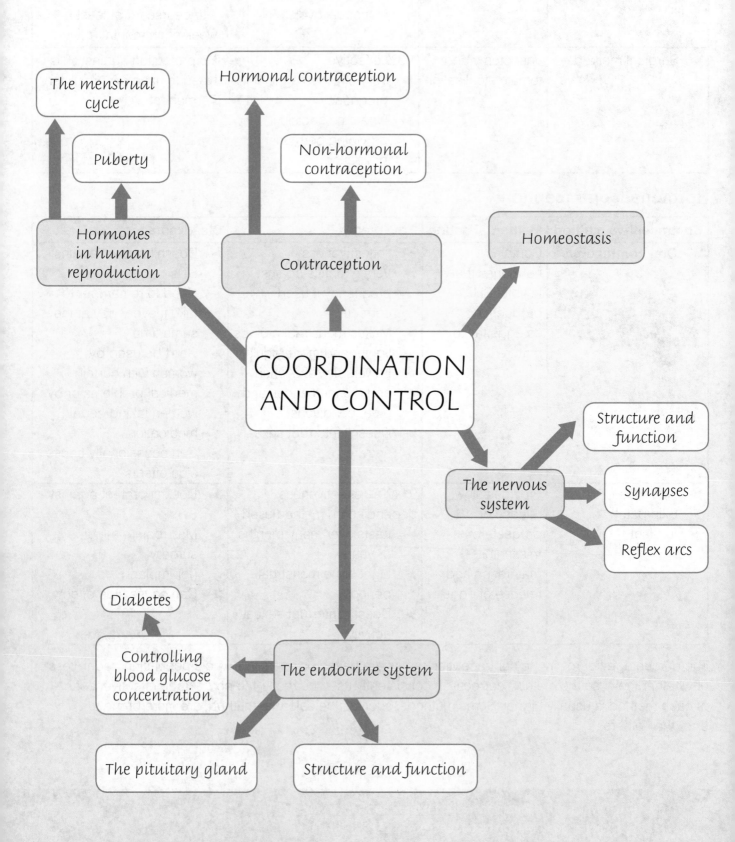

The menstrual cycle

Hormonal contraception

Puberty

Non-hormonal contraception

Hormones in human reproduction

Contraception

Homeostasis

COORDINATION AND CONTROL

Structure and function

The nervous system

Synapses

Reflex arcs

Diabetes

Controlling blood glucose concentration

The endocrine system

The pituitary gland

Structure and function

Practice questions

1. The flow chart shows the events that occur during a reflex action.

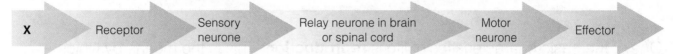

X → Receptor → Sensory neurone → Relay neurone in brain or spinal cord → Motor neurone → Effector

Paul accidentally puts his hand on a pin. Without thinking, he immediately pulls his hand away.

   a) Which component of a reflex arc is represented by the letter X?   **(1 mark)**

   b) Give **two** reasons why this can be described as a reflex action.   **(2 marks)**

   c) Use the flow chart to describe what happens in this reflex action.   **(4 marks)**

2. Look at the graph showing a person's blood sugar levels.

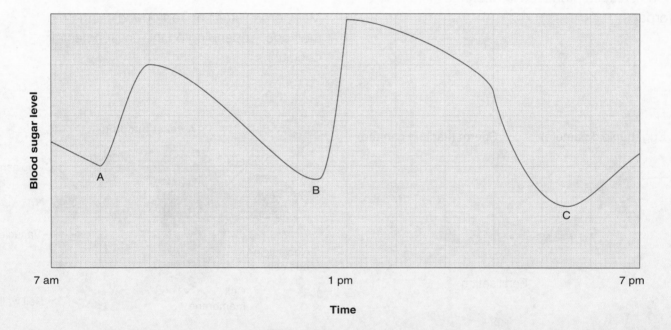

   a) How can we tell from the graph that this person has diabetes?   **(2 marks)**

   b) Explain why the person's blood sugar level rises rapidly just after points A and B.   **(1 mark)**

   c) Describe what would happen after points A and B if the person did not have diabetes.   **(1 mark)**

   d) Explain why the person needed to eat a chocolate at point C.   **(1 mark)**

Sexual and asexual reproduction

One of the basic characteristics of life is **reproduction**. This is the means by which a species continues. If sufficient offspring are not produced, the species becomes **extinct**.

Sexual reproduction

Sexual reproduction is where a male **gamete** (e.g. sperm) meets a female gamete (e.g. egg). This **fusion** of the two gametes is called **fertilisation** and may be **internal** or **external**.

Gametes are produced by **meiosis** in the sex organs.

Asexual reproduction

Asexual reproduction does not require different male and female cells. Instead, genetically identical clones are produced from mitosis. These may just be individual cells, as in the case of yeast, or whole multicellular organisms, e.g. **aphids**.

Many organisms can reproduce using both methods, depending on the environmental conditions.

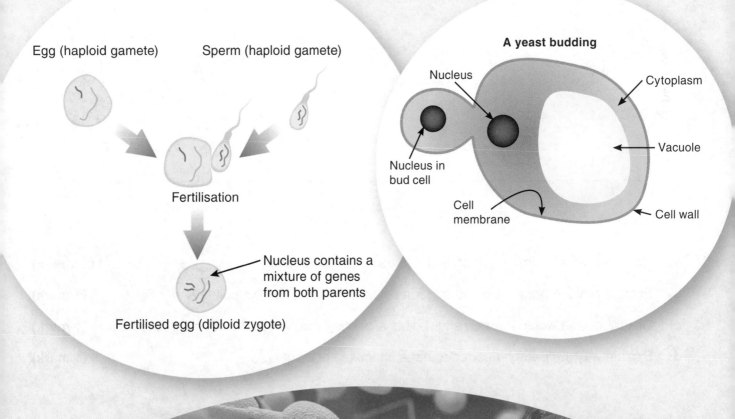

Egg (haploid gamete)   Sperm (haploid gamete)

Fertilisation

Nucleus contains a mixture of genes from both parents

Fertilised egg (diploid zygote)

A yeast budding

Nucleus

Cytoplasm

Nucleus in bud cell

Vacuole

Cell membrane

Cell wall

Comparing sexual and asexual reproduction

Advantages of sexual reproduction

- ➤ Produces **variation** in offspring through the process of meiosis, where genes are 'shuffled.' Variation is increased by **random fusion of gametes**.
- ➤ Survival advantage gained when the environment changes because different genetic types have more chance of producing well-adapted offspring.
- ➤ Humans can make use of sexual reproduction through **selective breeding**. This enhances food production.

Disadvantages of sexual reproduction

- ➤ Relatively slow process.
- ➤ Variation can be a disadvantage in stable environments.
- ➤ More resources required than for asexual reproduction, e.g. energy, time.
- ➤ Results of selective breeding are unpredictable and might lead to genetic abnormalities from 'in-breeding'.

Advantages of asexual reproduction

- ➤ Only one parent required.
- ➤ Fewer resources (energy and time) need to be devoted to finding a mate.
- ➤ Faster than sexual reproduction – survival advantage of producing many offspring in a short period of time.
- ➤ Many identical offspring of a well-adapted individual can be produced to take advantage of favourable conditions.

Disadvantages of asexual reproduction

- ➤ Offspring may not be well adapted in a changing environment.

Keywords

Fusion ➤ Joining together; in biology the term is used to describe fertilisation

Internal fertilisation ➤ Gametes join **inside** the body of the female

External fertilisation ➤ Gametes join **outside** the body of the female

Aphid ➤ A type of sap-sucking insect

Asexual reproduction

Sexual reproduction

1. Sexual reproduction requires a greater devotion of resources by an organism. So why do so many organisms use it?
2. State one advantage of asexual reproduction.

DNA

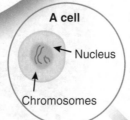

A cell

Nucleus

Chromosomes

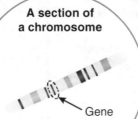

A section of a chromosome

Gene

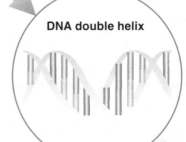

DNA double helix

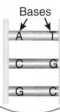

A section of the double helix

Bases

A — T

C — G

G — C

DNA and the genome

The nucleus of each cell contains a complete set of genetic instructions called the **genetic code**. The information is carried as genes, which are small sections of DNA found on **chromosomes**. The genetic code controls cell activity and, consequently, characteristics of the whole organism.

DNA facts

➤ DNA is a **polymer**.

➤ It is made of two strands coiled around each other called a **double helix**.

➤ The genetic code is in the form of nitrogenous **bases**.

➤ Bases bond together in pairs forming **hydrogen bond** cross-links.

➤ The structure of DNA was discovered in 1953 by **James Watson** and **Francis Crick**, using experimental data from **Rosalind Franklin** and **Maurice Wilkins**.

➤ A single gene codes for a particular sequence of **amino acids**, which, in turn, makes up a single **protein**.

➤ Construct part of a DNA molecule using plasticine. Use different colours to represent bases, sugar and phosphate molecules. Use the model to illustrate how changes in the base sequence can result in new proteins. You could use different colours to represent different amino acids. Make appropriate shapes to represent each component. The diagrams used in this module are a good starting point.

➤ Use your model as a way of learning the names of the different components.

Keywords

Polymer ➤ A long chain molecule made up of individual units called **monomers**

Hydrogen bond ➤ A bond formed between hydrogen and oxygen atoms on different molecules close to each other

Anthropologists ➤ Scientists who study the human race and its evolution

The human genome

The **genome** of an organism is the entire genetic material present in an adult cell of an organism.

The Human Genome Project (HGP)

The HGP was an international study. Its purpose was to map the complete set of genes in the human body.

HGP scientists worked out the code of the human genome in three ways. They:

➤ determined the sequence of all the bases in the genome

➤ drew up maps showing the locations of the genes on chromosomes

➤ produced linkage maps that could be used for tracking inherited traits from generation to generation, e.g. for genetic diseases. This could then lead to targeted treatments for these conditions.

The results of the project, which involved collaboration between UK and US scientists, were published in 2003. Three billion base pairs were determined.

The mapping of the human genome has enabled **anthropologists** to work out historical human migration patterns. This has been achieved by collecting and analysing DNA samples from many people across the globe. The study is called the **Genographic Project**.

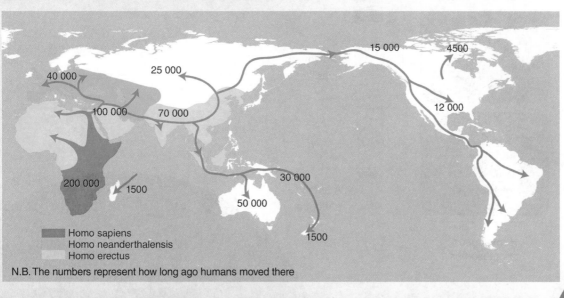

World map showing suggested migration patterns of early hominids

40 000
25 000
15 000
4500
100 000
70 000
12 000
200 000
1500
50 000
30 000
1500

- Homo sapiens
- Homo neanderthalensis
- Homo erectus

N.B. The numbers represent how long ago humans moved there

1. How has the Human Genome Project advanced medical science?
2. What do genes code for?

The genetic code

Monohybrid crosses

Most characteristics or **traits** are the result of multiple alleles interacting but some are controlled by a single gene. Examples include fur colour in mice and the shape of ear lobes in humans. These genes exist as pairs called **alleles** on **homologous chromosomes**.

Alleles are described as being **dominant** or **recessive**.

➤ A **dominant** allele controls the development of a characteristic even if it is present on only one chromosome in a pair.

➤ A **recessive** allele controls the development of a characteristic only if a dominant allele is not present, i.e. if the recessive allele is present on both chromosomes in a pair.

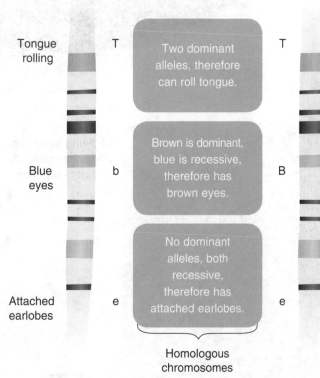

Tongue rolling	T	Two dominant alleles, therefore can roll tongue.	T
Blue eyes	b	Brown is dominant, blue is recessive, therefore has brown eyes.	B	Brown eyes
Attached earlobes	e	No dominant alleles, both recessive, therefore has attached earlobes.	e	Attached earlobes

Homologous chromosomes

Keywords

Alleles ➤ Alternative forms of a gene on a homologous pair of chromosomes
Homologous chromosomes ➤ A pair of chromosomes carrying alleles that code for the same characteristics

If **both chromosomes** in a pair contain the **same allele** of a gene, the individual is described as being **homozygous** for that gene or condition.

If the chromosomes in a pair contain **different alleles** of a gene, the individual is **heterozygous** for that gene or condition.

The combination of alleles for a particular characteristic is called the **genotype**. For example, the genotype for a homozygous dominant tongue-roller would be **TT**. The fact that this individual is able to roll their tongue is termed their **phenotype**.

Other examples are:
➤ **bb** (genotype), blue eyes (phenotype)
➤ **EE** or **Ee** (genotype), unattached/pendulous ear lobes (phenotype).

When a characteristic is determined by just one pair of alleles, as with eye colour and tongue rolling, it is called **monohybrid inheritance**.

During your course, you will be expected to recognise, draw and interpret scientific diagrams.

The way complementary strands of DNA are arranged can be worked out once you know that base T bonds with A and base C bonds with G.

Can you write out the complementary (DNA) strand to this sequence?

A T T A C G T G A G C C

The terms used in genetics can be confusing because many of the words sound like each other. Design a quiz or game that could help you memorise them.
For example, you could carry out a 'What am I?' quiz, where you come up with three clues of graded difficulty so that a friend can attempt to work out which word is being described.
➤ So, for **homozygous** your first clue could be the hardest and not give much information away, such as: 'I am a combination of genes'.
➤ If your partner gets it wrong, the next clue could be: 'The combinations can be dominant or recessive'.
➤ Finally, if they still don't provide a correct answer, the easiest clue could be: 'This word describes a genotype where both alleles are the same'.
There are lots of other possibilities for exploring vocabulary. By designing the quiz, the information becomes more securely embedded in your mind.

1. Describe the difference between a dominant and a recessive allele.
2. What is the difference between a gene and an allele?

Genetic diagrams

Genetic diagrams are used to show all the possible combinations of alleles and outcomes for a particular gene. They use:

➤ capital letters for dominant alleles
➤ lower-case letters for recessive alleles.

For eye colour, brown is dominant and blue is recessive.
So **B** represents a brown allele and **b** represents a blue allele.

Example 1

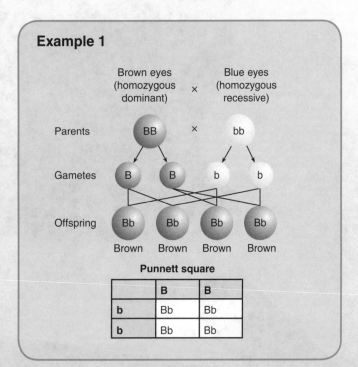

	B	B
b	Bb	Bb
b	Bb	Bb

Example 2

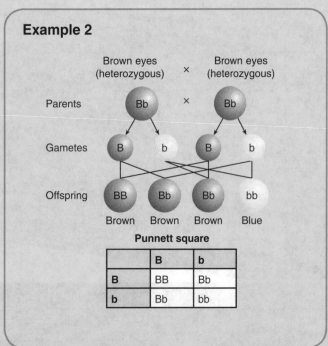

	B	b
B	BB	Bb
b	Bb	bb

Example 3

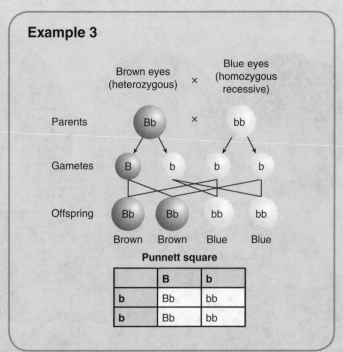

	B	b
b	Bb	bb
b	Bb	bb

WS You need to be able to interpret genetic diagrams and work out ratios of offspring.

➤ In Example 2 on the opposite page, the phenotypes of the offspring are 'brown eyes' and 'blue eyes'.
As there are potentially three times as many brown-eyed children as blue-eyed, the phenotypes are said to be in a 3:1 ratio.

➤ In Example 3 on the opposite page, the ratio would be 1:1 because half of the theoretical offspring are brown-eyed and half blue-eyed. Another way of saying this is that the probability of parents producing a brown-eyed child is 50%, or ½, or 0.5.

Most traits result not from one pair of alleles but from multiple genes interacting, e.g. inheritance of blood groups in the **ABO** system.

Family trees

Family trees are another way of showing how genetic traits can be passed on. Here is an example.

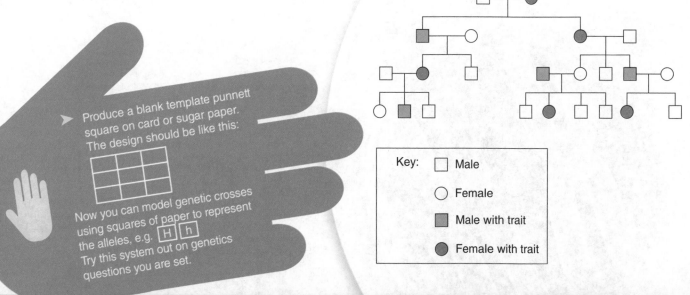

Key:
☐ Male
○ Female
■ Male with trait
● Female with trait

➤ Produce a blank template punnett square on card or sugar paper. The design should be like this:

Now you can model genetic crosses using squares of paper to represent the alleles, e.g. H h
Try this system out on genetics questions you are set.

1. If a homozygous brown-eyed individual is crossed with a homozygous blue-eyed individual, what is the probability of them producing a blue-eyed child?

Inheritance and genetic disorders 2

Inheritance of sex

Sex in humans/mammals is determined by whole chromosomes. These are the 23rd pair and are called sex chromosomes. There is an 'X' chromosome and a smaller 'Y' chromosome. The other 22 chromosome pairs carry the remainder of genes coding for the rest of the body's characteristics.

All egg cells carry X chromosomes. Half the sperm carry X chromosomes and half carry Y chromosomes. The sex of an individual depends on whether the egg is fertilised by an X-carrying sperm or a Y-carrying sperm.

If an X sperm fertilises the egg it will become a girl. If a Y sperm fertilises the egg it will become a boy. The chances of these events are equal, which results in approximately equal numbers of male and female offspring.

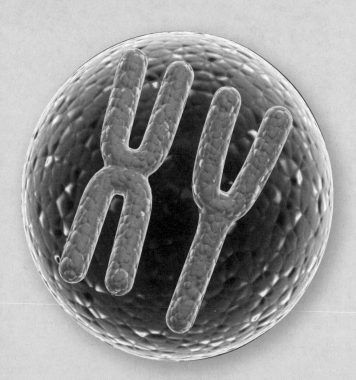

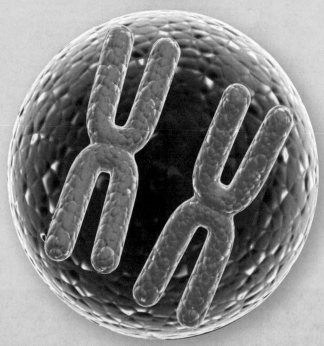

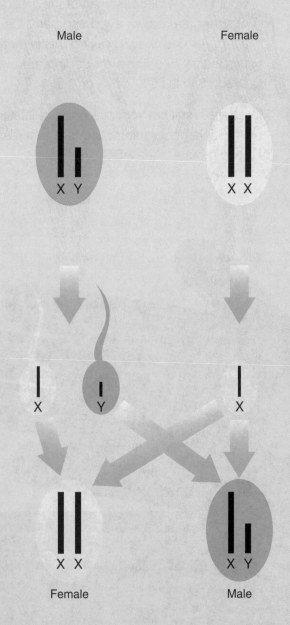

Male   Female

X Y   X X

X   Y   X

X X   X Y

Female   Male

Inherited diseases

Some disorders are caused by a 'faulty' gene, which means they can be **inherited**. One example is **polydactyly**, which is caused by a dominant allele and results in extra fingers or toes. The condition is not life-threatening.

Cystic fibrosis, on the other hand, can limit life expectancy. It causes the mucus in respiratory passages and the gut lining to be very thick, leading to build-up of phlegm and difficulty in producing correct digestive enzymes.

Cystic fibrosis is caused by a recessive allele. This means that an individual will only exhibit symptoms if both recessive alleles are present in the genotype. Those carrying just one allele will not show symptoms, but could potentially pass the condition on to offspring. Such people are called **carriers**.

Technology has advanced and it is now possible to screen embryos for genetic disorders.

➤ If an embryo has a life-threatening condition it could be destroyed, or simply not be implanted if IVF was being applied.

➤ Alternatively, new gene therapy techniques might be able to reverse the effects of the condition, resulting in a healthy baby.

As yet, these possibilities have to gain approval from ethics committees – some people think that this type of 'interference with nature' could have harmful consequences.

Conditions such as cystic fibrosis are mostly caused by **faulty alleles** that are **recessive**.

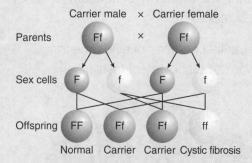

Knowing that there is a 1 in 4 chance that their child might have cystic fibrosis gives parents the opportunity to make decisions about whether to take the risk and have a child. This is a very difficult decision to make.

On a piece of plain paper, design a mind map to summarise the information you need to learn about genetic disorders. You could start by creating boxes for 'Cystic fibrosis', 'Polydactyly', 'Treatments' and 'How abnormalities occur'. Now you can populate these boxes with information. Produce branches to your diagram to allow for sub-categories for information, e.g. 'Symptoms of cystic fibrosis'.

---

1. What is the genotype of a human female?
2. How does the combination of two faulty, cystic fibrosis alleles affect the phenotype of the person who possesses them?

Variation and evolution

Keywords

Anatomy ➤ The study of structures within the bodies of organisms

Compress ➤ Squash or squeeze. In geology, this is usually due to Earth movements or laying down sediments

Extinction ➤ When there are no more members of a species left living

Variation

The two major factors that contribute to the appearance and function of an organism are:

➤ **genetic information** passed on from parents to offspring

➤ **environment** – the conditions that affect that organism during its lifetime, e.g. climate, diet, etc.

These two factors account for the large variation we see **within** and **between** species. In most cases, both of these factors play a part.

Evolution

Put simply, evolution is the theory that all organisms have arisen from simpler life forms over billions of years. It is driven by the **mechanism** of **natural selection**. For natural selection to occur, there must be genetic variation between individuals of the same species. This is caused by mutation or new combinations of genes resulting from sexual reproduction (see Module 25).

Most mutations have no effect on the phenotype of an organism. Where they do, and if the environment is changing, this can lead to relatively rapid change.

Evidence for evolution

Evidence for evolution comes from many sources. It includes:

➤ comparing genomes of different organisms

➤ studying embryos and their similarities

➤ looking at changes in species during modern times, e.g. antibiotic resistance in bacteria

➤ comparing the **anatomy** of different forms

➤ the fossil record.

One of the earliest sources of evidence for evolution was the discovery of fossils.

Use different-coloured pieces of card to depict one of the ideas covered in this module. Keep the design simple. For example, you could:

➤ produce five scenes showing the different stages of fossilisation.

When you have completed your design, explain the process/idea to a revision buddy.

How fossils are formed

1. When an animal or plant dies, the processes of decay usually cause all the body tissues to break down. In rare circumstances, the organism's body is rapidly **covered** and oxygen is prevented from reaching it. Instead of decay, fossilisation occurs.

2. Over hundreds of thousands of years, further sediments are laid down and **compress** the organism's remains.
3. Parts of the organism, such as bones and teeth, are **replaced by minerals** from solutions in the rock.

4. Earth upheavals, e.g. **tectonic plate movement**, bring sediments containing the fossils nearer the Earth's surface.

5. Erosion of the rock by wind, rain and rivers exposes the fossil. At this stage, the remains might be found and excavated by **paleontologists**.

Fossils can also be formed from footprints, burrows and traces of tree roots.

By comparing different fossils and where they are found in the rock layers, paleontologists can gain insights into how one form may have developed into another.

Difficulties occur with earlier life forms because many were **soft-bodied** and therefore not as well-preserved as organisms with bones or shells. Any that *are* formed are easily destroyed by Earth movements. As a result of this, scientists cannot be certain about exactly how life began.

Extinction

The fossil record provides evidence that most organisms that once existed have become **extinct**. In fact, there have been at least five **mass extinctions** in geological history where most organisms died out. One of these coincides with the disappearance of dinosaurs.

Causes of extinction include:
➤ **catastrophic events**, e.g. volcanic eruptions, asteroid collisions
➤ changes to the environment over geological time
➤ new **predators**
➤ new **diseases**
➤ new, more successful **competitors**.

1. State two pieces of evidence that support the theory of evolution through natural selection.
2. Why are fossils so rare?

Darwin and evolution

29

Keywords
Fittest ➤ The most adapted individual or species
Competition ➤ When two individuals or populations seek to exploit a resource, e.g. food. One individual/population will eventually replace the other

Darwin's theory of evolution through natural selection

Within a population of organisms there is a range of variation among individuals. This is caused by genes. Some differences will be beneficial; some will not.

Beneficial characteristics make an organism more likely to survive and pass on their genes to the next generation. This is especially true if the environment is changing. This ability to be successful is called **survival of the** fittest.

Species that are not well adapted to their environment may become extinct. This process of change is summed up in the theory of evolution through **natural selection**, put forward by **Charles Darwin** in the nineteenth century.

Charles Darwin

Many theories have tried to explain how life might have come about in its present form.

However, Darwin's theory is accepted by most scientists today. This is because it explains a wide range of observations and has been discussed and tested by many scientists.

Darwin's theory can be reduced to five ideas. They are:
➤ variation
➤ **competition**
➤ survival of the fittest
➤ inheritance
➤ extinction.

This activity should help you memorise Darwin's theory and learn how to apply the features using different scenarios.
➤ Arrange pieces of coloured paper into sets of five. About three lots will do – so fifteen altogether.
➤ On each set, write out the headings for the theory of evolution through natural selection: variation, competition, survival of the fittest, inheritance and extinction.
➤ Now research three case studies relating to natural selection, e.g. warfarin resistance in rats or the shape of shells in Galapagos tortoises. As you read, write down information on each of the five cards.

Darwin's ideas are illustrated in the following two examples.

Example 1: peppered moths

Variation – most peppered moths are pale and speckled. They are easily camouflaged amongst the lichens on silver birch tree bark. There are some rare, dark-coloured varieties (that originally arose from genetic mutation). They are easily seen and eaten by birds.

Competition – in areas with high levels of air pollution, lichens die and the bark becomes discoloured by soot. The lighter peppered moths are now put at a competitive disadvantage.

Survival of the fittest – the dark (melanic) moths are now more likely to avoid detection by predators.

Inheritance – the genes for dark colour are passed on to offspring and gradually become more common in the general population.

Extinction – if the environment remains polluted, the lighter form is more likely to become extinct.

Dark peppered moth

Peppered moth

Example 2: methicillin-resistant bacteria

The resistance of some bacteria to antibiotics is an increasing problem. MRSA bacteria have become more common in hospital wards and are difficult to eradicate.

Variation – bacteria mutate by chance, giving them a resistance to antibiotics.

Competition – the non-resistant bacteria are more likely to be killed by the antibiotic and become less competitive.

Survival of the fittest – the antibiotic-resistant bacteria survive and reproduce more often.

Inheritance – resistant bacteria pass on their genes to a new generation; the gene becomes more common in the general population.

Extinction – non-resistant bacteria are replaced by the newer, resistant strain.

To slow down the rate at which new, resistant strains of bacteria can develop:
➤ doctors are urged not to prescribe antibiotics for obvious viral infections or for mild bacterial infections
➤ patients should complete the full course of antibiotics to ensure that **all** bacteria are destroyed (so that none will survive to mutate into resistant strains).

Developing new antibiotics requires a lot of money and time and it is unlikely that this rate of development will be able to keep up with the speed at which resistant bacteria evolve.

Species become more and more specialised as they evolve and adapt to their environmental conditions.

The point at which a new species is formed occurs when the original population can no longer interbreed with the newer, 'mutant' population. For this to occur, **isolation** needs to happen.

Speciation

➤ Groups of the same species that are separated from each other by physical boundaries (like mountains or seas) will not be able to breed and share their genes. This is called **geographical isolation**.
➤ Over long periods of time, separate groups may specialise so much that they cannot successfully breed any longer and so two new species are formed – this is **reproductive isolation**.

1. Define evolution and natural selection.
2. State two examples where natural selection has been observed by scientists in recent times.
3. What has to happen to the beneficial genes for a helpful phenotype to spread through a population?

Selective breeding

Selective breeding

Farmers and dog breeders have used the principles of selective breeding for thousands of years by keeping the best animals and plants for breeding.

For example, to breed Dalmatian dogs, the spottiest dogs have been bred through the generations to eventually get Dalmatians. The factor most affected by selective breeding in dogs is probably temperament. Most breeds are either naturally obedient to humans or are trained to be so.

This is the process of selective breeding.

Select the desired characteristics in parents. → Allow the individuals to breed (or cross-pollinate if you are dealing with plants). → Select the desired offspring and allow them to become parents of the next generation.

This process has to be repeated many times to get the desired results.

Advantages of selective breeding
➤ It results in an organism with the 'right' characteristics for a particular function.
➤ In farming and horticulture, it is a more efficient and economically viable process than natural selection.

Disadvantages of selective breeding
➤ Intensive selective breeding reduces the gene pool – the range of alleles in the population decreases so there is **less variation**.
➤ Lower variation reduces a species' ability to respond to environmental change.
➤ It can lead to an accumulation of harmful recessive characteristics (in-breeding), e.g. restriction of respiratory pathways and dislocatable joints in bulldogs.

1. Write down one similarity and one difference between natural selection and selective breeding.

Examples of selective breeding
Modern food plants

Three of our modern vegetables have come from a single ancestor by selective breeding. (Remember, it can take many, many generations to get the desired results.)

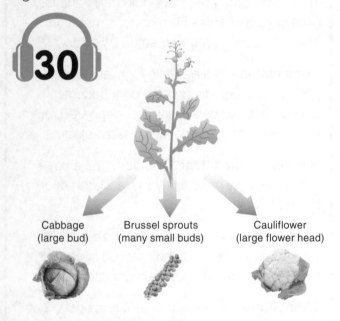

Cabbage (large bud) Brussel sprouts (many small buds) Cauliflower (large flower head)

Selective breeding in plants has also been undertaken to produce:
➤ disease resistance in crops
➤ large, unusual flowers in garden plants.

Modern cattle

Selective breeding can contribute to improved yields in cattle. Here are some examples.
➤ **Quantity of milk** – years of selecting and breeding cattle that produce larger than average quantities of milk has produced herds of cows that produce high daily volumes of milk.
➤ **Quality of milk** – as a result of selective breeding, Jersey cows produce milk that is rich and creamy, and can therefore be sold at a higher price.
➤ **Beef production** – the characteristics of the Hereford and Angus varieties have been selected for beef production over the past 200 years or more. They include hardiness, early maturity, high numbers of offspring and the swift, efficient conversion of grass into body mass (meat).

Design a poster describing the stages of selective breeding.

Genetic engineering

Keywords

Herbicide ➤ A chemical applied to crops to kill weeds

Genetic engineering

All living organisms use the same basic genetic code (DNA). So genes can be transferred from one organism to another in order to deliberately change the recipient's characteristics. This process is called genetic engineering or genetic modification (GM).

Altering the genetic make-up of an organism can be done for many reasons.

➤ **To improve resistance to herbicides**: for example, soya plants are genetically modified by inserting a gene that makes them resistant to a herbicide. When the crop fields are sprayed with the herbicide only the weeds die, leaving the soya plants without competition so they can grow better. Resistance to frost or disease can also be genetically engineered. Bigger yields result.

➤ **To improve the quality of food**: for example, bigger and more tasty fruit.

➤ **To produce a substance you require**: for example, the gene for human insulin can be inserted into bacteria or fungi, to make human insulin on a large scale to treat diabetes.

➤ **Disease resistance**: crop plants receive genes that give them resistance to the bacterium *Bacillus thuringiensis*.

Advantages of genetic engineering
➤ It allows organisms with new features to be produced rapidly.
➤ It can be used to make biochemical processes cheaper and more efficient.
➤ In the future, it may be possible to use genetic engineering to change a person's genes and cure certain disorders, e.g. cystic fibrosis. This is an area of research called gene therapy.

Disadvantages of genetic engineering
➤ Transplanted genes may have unexpected harmful effects on human health.
➤ Some people are worried that GM plants may cross-breed with wild plants and release their new genes into the environment.

1. What advantages does genetic engineering have over selective breeding?
2. Should we be expanding the range of GM foods we eat? Give one reason **for** this proposal and one reason **against**.

There is a huge variety of living organisms. Scientists group or classify them using shared characteristics. This is important because it helps to:

➤ work out how organisms evolved on Earth
➤ understand how organisms coexist in ecological communities
➤ identify and monitor rare organisms that are at risk from extinction.

Classification

Carl Linnaeus

The origins of classification

In the past, observable characteristics were used to place organisms into categories.

In the eighteenth century, **Carl Linnaeus** produced the first classification system. He developed a hierarchical arrangement where larger groups were subdivided into smaller ones.

Kingdom	Largest group
Phylum	
Class	
Order	
Family	
Genus	
Species	Smallest group

Linnaeus also developed a **binomial system** for naming organisms according to their genus and species. For example, the common domestic cat is *Felis catus*. Its full classification would be:

➤ Kingdom: *Animalia*
➤ Phylum: *Chordata*
➤ Class: *Mammalia*
➤ Order: *Carnivora*
➤ Family: *Felidae*
➤ Genus: *Felis*
➤ Species: *Catus*.

Linnaeus' system was built on and resulted in a **five-kingdom system**. Developments that contributed to the introduction of this system included improvements in microscopes and a more thorough understanding of the biochemical processes that occur in all living things. For example, the presence of particular chemical pathways in a range of organisms indicated that they probably had a **common ancestor** and so were more closely related than organisms that didn't share these pathways.

Kingdom	Features	Examples
Plants	Cellulose cell wall Use light energy to produce food	Flowering plants Trees Ferns Mosses
Animals	Multicellular Feed on other organisms	Vertebrates Invertebrates
Fungi	Cell wall of chitin Produce spores	Toadstools Mushrooms Yeasts Moulds
Protoctista Protozoa	Mostly single-celled organisms	Amoeba Paramecium
Prokaryotes	No nucleus	Bacteria Blue–green algae

Keywords

Binomial system ➤ 'Two name' system of naming species using Latin

Common ancestor ➤ Organism that gave rise to two different branches of organisms in an evolutionary tree

The classification diagram below illustrates how different lines of evidence can be used. The classes of vertebrates share a common ancestor and so are quite closely related. Evidence for this lies in comparative anatomy and similarities in biochemical pathways.

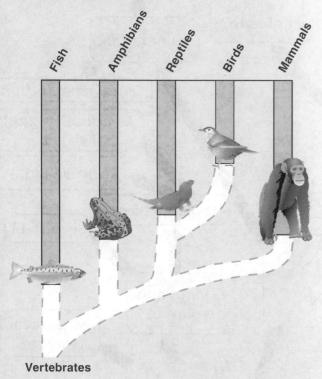

Vertebrates

In more recent times, improvements in science have led to a **three-domain system** developed by **Carl Woese**. In this system organisms are split into:

➤ **archaea** (primitive bacteria)
➤ **bacteria** (true bacteria)
➤ **eukaryota** (including Protista, fungi, plants and animals).

Further improvements in science include chemical analysis and further refinements in comparisons between non-coding sections of DNA.

Evolutionary trees

Tree diagrams are useful for depicting relationships between similar groups of organisms and determining how they may have developed from common ancestors. Fossil evidence can be invaluable in establishing these relationships.

Here, two species are shown to have evolved from a common ancestor.

Common ancestor

WS You need to understand how new evidence and data leads to changes in models and theories.

In the case of the three-domain system, a more accurate and cohesive classification structure was proposed as a result of improvements in microscopy and increasing knowledge of organisms' internal structures.

These apes share a common ancestor

1. What do the first and second words in the binomial name of an organism mean?
2. Name four groups of organisms found in the domain *Eukaryota*.
3. It is thought that chimpanzees and humans evolved from a now extinct ape. What term describes this animal?

32

Mind map

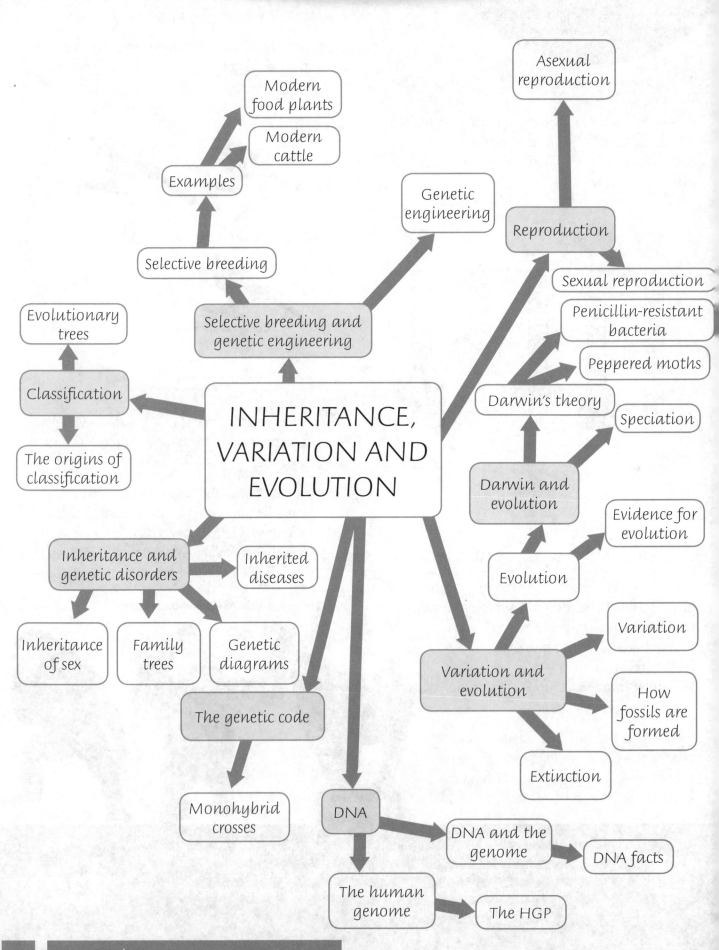

Modern food plants

Modern cattle

Examples

Genetic engineering

Asexual reproduction

Reproduction

Selective breeding

Sexual reproduction

Evolutionary trees

Selective breeding and genetic engineering

Penicillin-resistant bacteria

Peppered moths

Classification

Darwin's theory

Speciation

INHERITANCE, VARIATION AND EVOLUTION

The origins of classification

Darwin and evolution

Evidence for evolution

Evolution

Inheritance and genetic disorders

Inherited diseases

Variation

Inheritance of sex

Family trees

Genetic diagrams

Variation and evolution

How fossils are formed

The genetic code

Extinction

Monohybrid crosses

DNA

DNA and the genome

DNA facts

The human genome

The HGP

Practice questions

1. Scientists believe that the whale may have evolved from a horse-like ancestor that lived in swampy regions millions of years ago. Suggest how whales could have evolved from a horse-like mammal. In your answer, use Darwin's theory of natural selection. **(4 marks)**

2. The diagram below shows the inheritance of cystic fibrosis in a family.

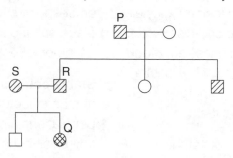

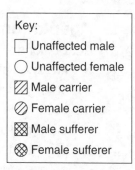

Cystic fibrosis is caused by a recessive allele, f. The dominant allele of the gene is represented by F.

a) Give the alleles for person P. **(1 mark)**

b) Give the alleles for person Q. **(1 mark)**

3. The diagram shows a molecule of DNA.

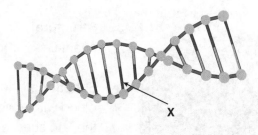

a) Name the part labelled **X** in the DNA diagram. **(1 mark)**

b) The 'backbone' of the molecule is arranged in such a way as to make it very stable. Name the term used to describe the shape of DNA. **(1 mark)**

Organisms and ecosystems

33

Keyword

Environmental resources ➤ Materials or factors that organisms need to survive, e.g. high oxygen concentration, living space or a particular food supply

Communities

Ecosystems are physical environments with a particular set of conditions (**abiotic** factors), plus all the organisms that live in them. The organisms interact through competition and predation (these are **biotic** factors). An ecosystem can support itself without any influx of other factors or materials. Its energy source (usually the Sun) is the only external factor.

Other terms help to describe aspects of the environment.

➤ The **habitat** of an animal or plant is the part of the physical environment where it lives. There are many types of habitat, each with particular characteristics, e.g. pond, hedgerow, coral reef.

➤ A **population** is the number of individuals of a species in a defined area.

➤ A **community** is the total number of individuals of all the different populations of organisms that live together in a habitat at any one time.

An organism must be well-suited to its habitat to be able to compete with other species for limited **environmental resources**. Even organisms within the same species may compete in order to survive and breed. Organisms that are specialised in this way are restricted to that type of habitat because their adaptations are unsuitable elsewhere.

Resources that plants compete over include:
➤ light
➤ space
➤ water
➤ minerals.

Animals compete over:
➤ food
➤ mates.
➤ territory.

Interdependence

In communities, each species may depend on other species for food, shelter, pollination and seed dispersal. If one species is removed, it may have knock-on effects for other species.

Stable communities contain species whose numbers fluctuate very little over time. The populations are in balance with the physical factors that exist in that habitat. Stable communities include **tropical rainforests** and ancient oak **woodlands**.

Adaptations

Adaptations:

➤ are special features or behaviours that make an organism particularly well-suited to its environment and better able to compete with other organisms for limited resources

➤ can be thought of as a biological solution to an environmental challenge – evolution provides the solution and makes species fit their environment.

Animals have developed in many different ways to become well adapted to their environment and to help them survive. Adaptations are usually of three types:

➤ **Structural** – for example, skin colouration in chameleons provides camouflage to hide them from predators.

➤ **Functional** – for example, some worms have blood with a high affinity for oxygen; this helps them to survive in anaerobic environments.

➤ **Behavioural** – for example, penguins huddle together to conserve body heat in the Antarctic habitat.

Look at the **polar bear** and its life in a very cold climate. It has:

➤ small ears and large bulk to reduce its surface area to volume ratio and so reduce heat loss

➤ a large amount of insulating fat (blubber)

➤ thick white fur for insulation and camouflage

➤ large feet to spread its weight on snow and ice

➤ fur on the soles of its paws for insulation and grip

➤ powerful legs so it is a good swimmer and runner, which enables it to catch its food

➤ sharp claws and teeth to capture prey.

The **cactus** is well adapted to living in a desert habitat. It:

➤ has a rounded shape, which gives a small surface area to volume ratio and therefore reduces water loss

➤ has a thick waxy cuticle to reduce water loss

➤ stores water in a spongy layer inside its stem to resist drought

➤ has sunken stomata, meaning that air movement is reduced, minimising loss of water vapour through them

➤ has leaves that take the form of spines to reduce water loss and to protect the cactus from predators.

Some organisms have biochemical adaptations. **Extremophiles** can survive extreme environmental conditions. For example:

➤ bacteria living in deep sea vents have optimum temperatures for enzymes that are much higher than 37°C

➤ icefish have antifreeze chemicals in their bodies, which lower the freezing point of body fluids

➤ some organisms can resist high salt concentrations or pressure.

WS During your course you will be asked to suggest explanations for observations made in the field or laboratory. These include:

➤ suggesting factors for which organisms are competing in a certain habitat

➤ giving possible adaptations for organisms in a habitat.

For example, low-lying plants in forest ecosystems often have specific adaptations for maximising light absorption as they are shaded by taller plants. Adaptations might include leaves with a large surface area and higher concentrations of photosynthetic pigments to absorb the correct wavelengths and lower intensities of light.

1. How is an ecosystem different from a habitat?
2. Which is the more stable community – a mixed-leaf woodland or a dry river bed in Africa? What is the reason for this?
3. How is a community different from a population?
4. Give two examples of extremophiles.
5. Why is it important for organisms to be well adapted?

Studying ecosystems

Testing soil pH

Taking measurements in ecosystems

Ecosystems involve the interaction between **non-living (abiotic)** and **living (biotic)** parts of the environment. So it is important to identify which factors need to be measured in a particular habitat.

Abiotic factors include:
➤ light intensity
➤ temperature
➤ moisture levels
➤ soil pH and mineral content
➤ wind intensity and direction
➤ carbon dioxide levels for plants
➤ oxygen levels for aquatic animals.

Biotic factors include:
➤ availability of food
➤ new predators arriving
➤ new pathogens
➤ one species out-competing another.

Measuring biotic factors – sampling methods

It is usually impossible to count all the species living in a particular area, so a **sample** is taken.

When sampling, make sure you:
➤ **take a big enough sample** to make the estimate good and reliable – the larger the sample, the more accurate the results.
➤ **sample randomly** – the more random the sample, the more likely it is to be representative of the population.

Quadrats

Quadrats are square frames that typically have sides of length 0.5 m. They provide excellent results as long as they are placed randomly. The population of a certain species can then be estimated.

For example, if an average of 4 dandelion plants are found in a 0.25 m² quadrat, a scientist would estimate that 16 dandelion plants would be found in each 1 m² and 16 000 dandelion plants in a 1000 m² field.

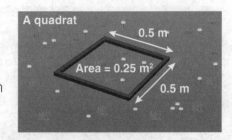

A quadrat
0.5 m
Area = 0.25 m²
0.5 m

Transects

Sometimes an environmental scientist may want to look at how species change across a habitat, or the boundary between two different habitats – for example, the plants found in a field as you move away from a hedgerow.

This needs a different approach that is systematic rather than random.

1. Lay down a line such as a tape measure. Mark regular intervals on it.
2. Next to the line, lay down a small quadrat. Estimate or count the number of plants of the different species. This can sometimes be done by estimating the percentage cover.
3. Move the quadrat along at regular intervals. Estimate and record the plant populations at each point until the end of the line.

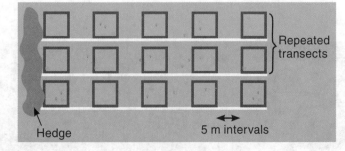

Repeated transects

Hedge   5 m intervals

Sampling methods

Sampling animal populations is more problematic as they are mobile and well adapted to evade capture. Here are three of the main techniques used.

Keyword

Sample ➤ A small area or population that is measured to give an indication of numbers within a larger area or population

Pooters 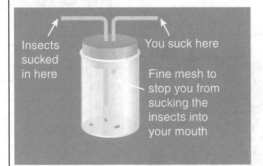	This is a simple technique in which insects are gathered up easily without harm. With this method, you get to find out which species are actually present, although you have to be systematic about your sampling in order to get representative results and it is difficult to get ideas of numbers.
Sweepnets 	Sweepnets are used in long grass or moderately dense woodland where there are lots of shrubs. Again, it is difficult to get truly representative samples, particularly in terms of the relative numbers of organisms.
Pitfall traps 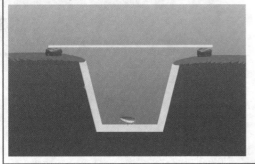	Pitfall traps are set into the ground and used to catch small insects, e.g. beetles. Sometimes a mixture of ethanol or detergent and water is placed in the bottom of the trap to kill the samples, and prevent them from escaping. This method can give an indication of the relative numbers of organisms in a given area if enough traps are used to give a representative sample.

Design a poster showing the different sampling methods.

1. In what situation would you use a transect? What information would it give you?

Feeding relationships

Predator–Prey relationships

Animals that kill and eat other animals are called **predators** (e.g. foxes, lynx). The animals that are eaten are called **prey** (e.g. rabbits, snowshoe hares).

Many animals can be both predator and prey. For instance, a stoat is a predator when it hunts rabbits and it is the prey when it is hunted by a fox.

Predator – stoat

Predator – fox

Prey – rabbit

Prey – stoat

In nature there is a delicate balance between the population of a predator (e.g. lynx) and its prey (e.g. snowshoe hare). However, the prey will always outnumber the predators.

The number of predators and prey follow a classic population cycle. There will always be more hares than lynx and the population peak for the lynx will always come after the population peak for the hare. As the population cycle is cause and effect, they will always be out of phase.

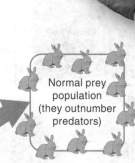

Normal prey population (they outnumber predators)

Predator population increases as plenty of food is available

Decrease in prey population as more are being eaten by increased number of predators

Decrease in predator population as there is now not enough food

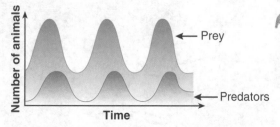

Trophic levels

Communities of organisms are organised in an ecosystem according to their feeding habits.

Food chains show:
- the organisms that consume other organisms
- the transfer of **energy** and **materials** from organism to organism.

Energy from the Sun enters most food chains when green plants absorb sunlight to **photosynthesise**. Photosynthetic and chemosynthetic organisms are the producers of **biomass** for the Earth. Feeding passes this energy and biomass from one organism to the next along the food chain.

A food chain

Green plant:
producer

Rabbit:
primary consumer

Stoat:
secondary consumer

Fox:
tertiary consumer

The arrow shows the flow of energy and biomass along the food chain.
- All food chains start with a **producer**.
- The rabbit is a herbivore (plant eater), also known as the **primary consumer**.
- The stoat is a carnivore (meat eater), also known as the **secondary consumer**.
- The fox is the top carnivore in this food chain, the **tertiary consumer**.

Each consumer or producer occupies a **trophic level** (feeding level).
- Level 1 are producers.
- Level 2 are primary consumers.
- Level 3 are secondary consumers.
- Level 4 are tertiary consumers.

Excretory products and uneaten parts of organisms can be the starting points for other food chains, especially those involving **decomposers**.

Decomposers

Keyword

Decomposers ➤ Microorganisms that break down dead plant and animal material by secreting enzymes into the environment. Small, soluble molecules can then be absorbed back into the microorganism by the process of diffusion

To demonstrate predator/prey cycles, do the following.
- Cut out multiple pictures of the animals in a particular predator–prey relationship (e.g. foxes and rabbits).
- Show these animals grouped together – try 5 foxes and 15 rabbits.
- Add 3 foxes. How will this affect the number of rabbits?
- Explain this to a revision buddy.
- Now show the changes in number of both rabbits and foxes as time goes by. Ask your buddy to rate your explanation.
- Swap roles.

1. What is meant by a trophic level?
2. Explain how a population of foxes might rise and fall with a population of rabbits.

Environmental change

Environmental change

Waste management

The human population is increasing exponentially (i.e. at a rapidly increasing rate). This is because birth rates exceed death rates by a large margin.

So the use of finite resources like fossil fuels and minerals is accelerating. In addition, waste production is going up:

➤ **on land**, from domestic waste in landfill, toxic chemical waste, **pesticides** and **herbicides**
➤ **in water**, from sewage fertiliser and toxic chemicals
➤ **in the air**, from smoke, carbon dioxide and sulfur dioxide.

Acid rain

When coal or oils are burned, sulfur dioxide is produced. Sulfur dioxide and nitrogen dioxide dissolve in water to produce acid rain.

Acid rain can:

➤ damage trees, stonework and metals
➤ make rivers and lakes acidic, which means some organisms can no longer survive.

The acids can be carried a long way away from the factories where they are produced. Acid rain falling in one country could be the result of fossil fuels being burned in another country

Keywords

Pesticide ➤ Chemical sprayed on crops to kill invertebrate pests

Herbicide ➤ Chemical sprayed on crops to kill weeds

Carbon sinks ➤ Resources that lock up carbon in their structure rather than allowing them to form carbon dioxide, e.g. peat bogs, oceans, limestone deposits

Endangered ➤ Category of risk attached to rare species of plants and animals. This usually triggers efforts to preserve the species' numbers

Emissions ➤ Gaseous products usually connected with pollution, e.g. carbon dioxide emissions from exhausts

Regeneration ➤ Rebuilding or regrowth of a habitat, e.g. flooding peatland to encourage regrowth of mosses and other plants

The greenhouse effect and global warming

The diagram explains how global warming can lead to climate change. This in turn leads to lower biodiversity.

Small amount of infrared radiation transmitted to space

CO_2 and CH_4 in the atmosphere absorb some of the energy and radiate it back to Earth

Rays from the Sun reach Earth and are reflected back towards the atmosphere

The consequences of global warming are:

➤ a rise in sea levels leading to flooding in low-lying areas and loss of habitat

➤ the migration of species and changes in their distribution due to more extreme temperature and rainfall patterns; some organisms won't survive being displaced into new habitats, or newly migrated species may outcompete native species. The overall effect is a loss of biodiversity.

Create a board game called 'Conservation' that has 100 squares. The object of the game is to improve the quality and quantity of the world's ecosystems. The winner is the first person to reach square 100.

Here is an idea for a **bonus** square.
➤ Establish a breeding programme for endangered snow-leopards. (Throw the dice again.)

Here is an idea for a penalty square.
➤ Deforestation of Brazilian rainforest lowers biodiversity. (Go back three spaces.)

Try to grade bonuses and penalties according to their impact.

WS You may be asked to evaluate methods used to address problems caused by human impact on the environment.

For example, here are some figures relating to quotas and numbers of haddock in the North Sea in two successive years.

	2009	2010
Haddock quota (tonnes)	27 507	23 381
Estimated population (thousands)	102	101

What conclusions could you draw from this data? What additional information would you need to give a more accurate picture?

Human activity can cause significant changes in the environment including:
➤ Temperature
➤ Availability of water
➤ Composition of atmospheric gases

1. Name one pollutant gas that contributes to acid rain.
2. The human population is increasing exponentially – what does this term mean?

Biodiversity

Biodiversity

Biodiversity is a measure of the number and variety of species within an ecosystem. A healthy ecosystem:

➤ has a large biodiversity
➤ has a large degree of interdependence between species
➤ is stable.

Species depend on each other for food, shelter and keeping the external physical environment maintained. Humans have had a negative impact on biodiversity due to:

➤ pollution killing plants and animals
➤ degrading the environment through deforestation and removing resources such as minerals and fossil fuels
➤ over-exploiting habitats and organisms.

Only recently have humans made efforts to reduce their impact on the environment. It is recognised that maintaining biodiversity is important to ensure the continued survival of the human race.

Impact of land use

As humans increase their economic activity, they use land that would otherwise be inhabited by living organisms. Examples of habitat destruction include:

➤ farming
➤ quarrying
➤ dumping waste in landfill.

Peat bogs

Peat bogs are important habitats. They support a wide variety of organisms and act as **carbon sinks**.

If peat is burned it releases carbon dioxide into the atmosphere and contributes to global warming. Removing peat for use as compost in gardens takes away the habitat for specialised animals and plants that aren't found in other habitats.

A marble quarry

Peat cut and left to dry

Deforestation

Deforestation is a particular problem in tropical regions. Tropical rainforests are removed to:

➤ **release land for cattle and rice fields** – these are needed to feed the world's growing population and for increasingly Western-style diets

➤ **grow crops for biofuel** – the crops are converted to **ethanol-based** fuels for use in petrol and diesel engines. Some specialised engines can run off pure ethanol.

The consequences of deforestation are:

➤ There are fewer plants, particularly trees, to absorb carbon dioxide. This leads to increased carbon dioxide in the atmosphere and accelerated global warming.
➤ Combustion and decay of the wood from deforestation releases more carbon dioxide into the atmosphere.
➤ There is reduced biodiversity as animals lose their habitats and food sources.

Maintaining biodiversity

To prevent further losses in biodiversity and to improve the balance of ecosystems, scientists, the government and environmental organisations can take action.

➤ Scientists establish **breeding programmes** for **endangered species**. These may be captive methods where animals are enclosed, or protection schemes that allow rare species to breed without being poached or killed illegally.

➤ The government sets **limits** on **deforestation** and **greenhouse gas emissions**.

Environmental organisations:

➤ **protect and regenerate** shrinking habitats such as mangrove swamps, heathlands and coral reefs

➤ **conserve and replant** hedgerows around the margins of fields used for crop growth

➤ introduce and encourage **recycling** initiatives that reduce the volume of landfill.

1. Why is high biodiversity seen as a good thing?
2. What steps could you take to maintain the biodiversity of a British mixed woodland?

Recycling

Materials within ecosystems are constantly being recycled and used to provide the substances that make up future organisms. Two of these substances are water and carbon.

The water cycle

Water is a vital part of the **biosphere**. Most organisms consist of over 50% water.

The two key processes that drive the water cycle are **evaporation** and **condensation**.

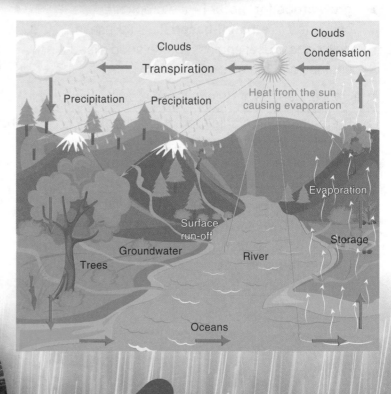

Produce a poster / flow diagram to illustrate the carbon cycle. Include all the organism types involved and the important processes.

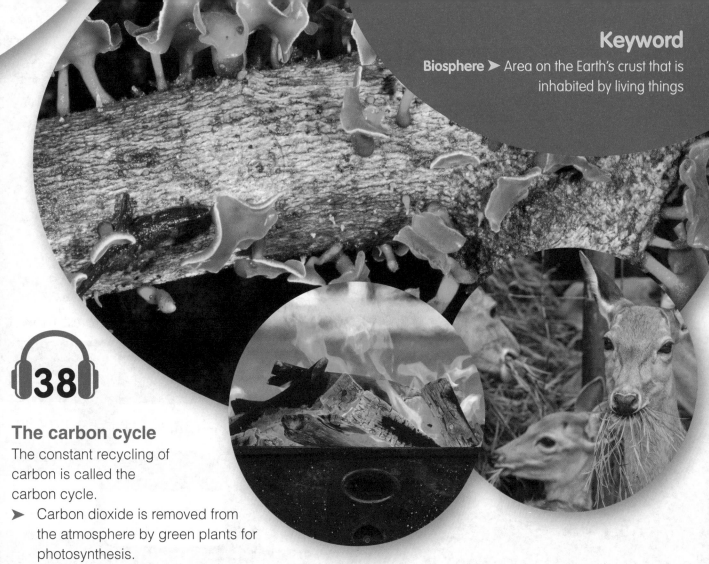

38

The carbon cycle

The constant recycling of carbon is called the carbon cycle.

➤ Carbon dioxide is removed from the atmosphere by green plants for photosynthesis.

➤ Plants and animals respire, releasing carbon dioxide into the atmosphere.

➤ Animals eat plants and other animals, which incorporates carbon into their bodies. In this way, carbon is passed along food chains and webs.

➤ Microorganisms such as fungi and bacteria feed on dead plants and animals, causing them to decay. The microorganisms respire and release carbon dioxide gas into the air. Mineral ions are returned to the soil through decay. This extraction and return of nutrients to the soil is called the decay cycle.

➤ Some organisms' bodies are turned into fossil fuels over millions of years, trapping the carbon as coal, peat, oil and gas.

➤ When fossil fuels are burned (combustion), the carbon dioxide is returned to the atmosphere.

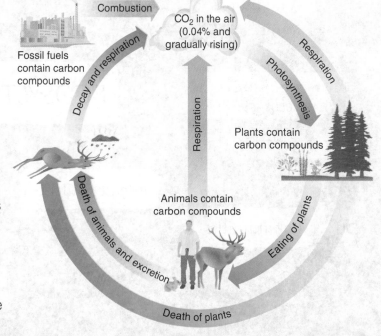

Combustion

Decay and respiration

Fossil fuels contain carbon compounds

CO_2 in the air (0.04% and gradually rising)

Respiration

Photosynthesis

Respiration

Plants contain carbon compounds

Death of animals and excretion

Animals contain carbon compounds

Eating of plants

Death of plants

1. Which two processes drive the water cycle?
2. How can carbon be stored in rocks?
3. Name two processes that add carbon dioxide to the atmosphere.

Mind map

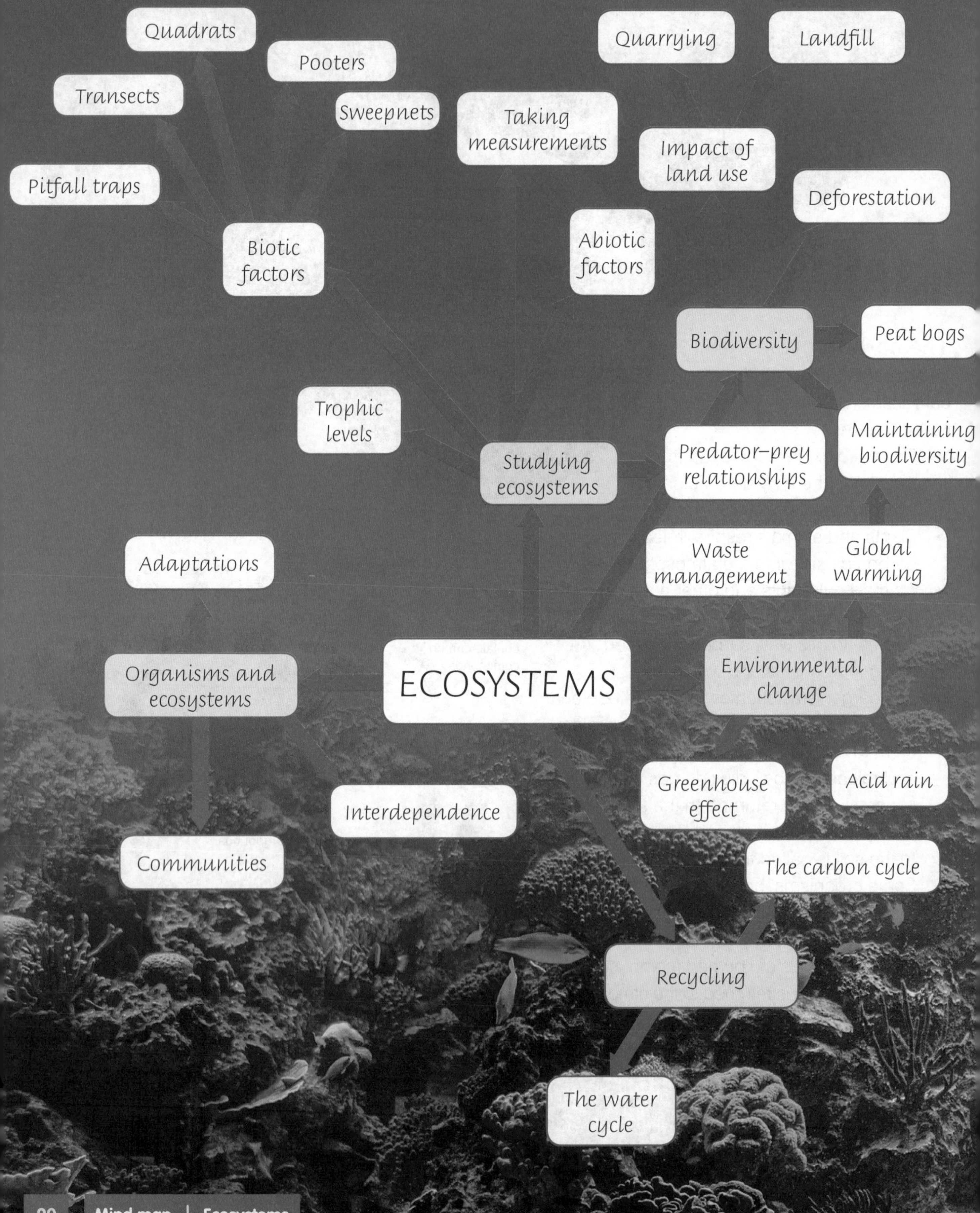

Quadrats

Pooters

Transects

Sweepnets

Pitfall traps

Quarrying

Landfill

Taking measurements

Impact of land use

Deforestation

Biotic factors

Abiotic factors

Biodiversity

Peat bogs

Trophic levels

Studying ecosystems

Predator–prey relationships

Maintaining biodiversity

Adaptations

Waste management

Global warming

Organisms and ecosystems

ECOSYSTEMS

Environmental change

Greenhouse effect

Acid rain

Interdependence

Communities

The carbon cycle

Recycling

The water cycle

Practice questions

1. Carbon is recycled in the environment in a process called the **carbon cycle**. The main processes of the carbon cycle are shown on the right.

   a) Name the process that occurs at stage **3** in the diagram. **(1 mark)**

   b) The UK government is planning to use fewer fossil-fuel-burning power stations in the future. How might this affect the carbon cycle? Use ideas about **combustion** and **fossil fuel formation** in your answer. **(2 marks)**

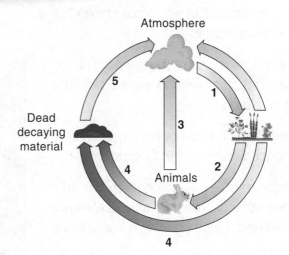

2. An environmental scientist observed and measured a kingfisher and fish population in a county's rivers over 10 years. She recorded her results as a graph.

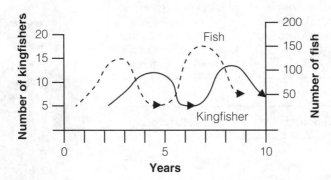

   a) How many fish were recorded in the third year? **(1 mark)**

   b) Describe how the size of the kingfisher population affected the size of the fish population. **(1 mark)**

   c) Describe the role that environmental organisations can play in conserving habitats. **(2 marks)**

3. Choose the correct words to complete the following passage about adaptations. **(3 marks)**

environment	population	features	community
characteristics	survival	evolutionary	predatory	suited

   Adaptations are special .... or .... that make a
   living organism particularly well .... to its ....
   Adaptations are part of an .... process that increases a living organism's
   chance of ....

Atoms and elements

➤ All substances are made of **atoms**.
➤ An atom is the smallest part of an **element** that can exist.
➤ There are approximately 100 different elements, all of which are shown in the **periodic table**.

On the right is part of the periodic table, showing the names and symbols of the first 20 elements.

Atoms of each element are represented by a chemical symbol in the periodic table.

Some elements have more than one letter in their symbol. The first letter is always a capital and the other letter is lower case.

Atoms and the periodic table

1 hydrogen **H**								2 helium **He**
3 lithium **Li**	4 beryllium **Be**	5 boron **B**	6 carbon **C**	7 nitrogen **N**	8 oxygen **O**	9 fluorine **F**	10 neon **Ne**	
11 sodium **Na**	12 magnesium **Mg**	13 aluminium **Al**	14 silicon **Si**	15 phosphorus **P**	16 sulfur **S**	17 chlorine **Cl**	18 argon **Ar**	
19 potassium **K**	20 calcium **Ca**							

The symbol for magnesium is **Mg**

The symbol for the element oxygen is **O**

Elements, compounds and mixtures

Elements contain only one type of atom and consist of single atoms or atoms bonded together	Compounds contain two (or more) elements that are chemically combined in fixed proportions	Mixtures contain two or more elements (or compounds) that are together but not chemically combined

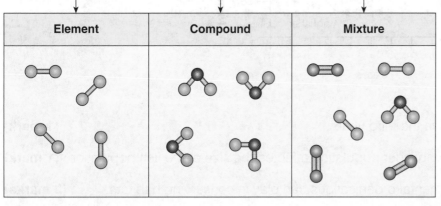

Element	Compound	Mixture

Chemical equations

We can use word or symbol equations to represent chemical reactions.

Word equation: sodium + oxygen → sodium oxide

Symbol equation: $4Na + O_2 \rightarrow 2Na_2O$

Write the name of each method of separating a mixture on some cards. Use a new card for each method. Write on other cards a description of each method of separation, again using a new card for each description. Shuffle the cards. Now try to match the names and descriptions.

Keywords

Atom ➤ The smallest part of an element that can enter into a chemical reaction
Element ➤ A substance that consists of only one type of atom
Compound ➤ A substance consisting of two or more different elements chemically combined together
Mixture ➤ Two or more elements or compounds that are not chemically combined

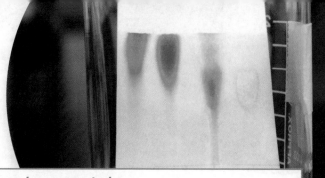

Separating mixtures

Compounds can only be separated by chemical reactions but **mixtures** can be separated by physical processes (not involving chemical reactions) such as filtration, crystallisation, simple (and fractional) distillation and chromatography.

How different mixtures can be separated		
Method of separation	**Diagram**	**Use**
Filtration	Filter paper Filter funnel Sand and water Sand Beaker Clear water (filtrate)	Used for separating insoluble solids from liquids, e.g. sand from water
Crystallisation	Evaporating dish Mixture Wire gauze Tripod stand Bunsen burner	Used to separate solids from solutions, e.g. obtaining salt from salty water
Simple distillation	Thermometer Water out Round-bottomed flask Liebig condenser Water in Heat	Simple distillation is used to separate a liquid from a solution, e.g. water from salty water
Fractional distillation		Fractional distillation is used to separate liquids that have different boiling points, e.g. ethanol and water
Chromatography	Solvent front Separated dyes Filter paper Ink spots Pencil line Solvent Spot of mixture	Used to separate dyes, e.g. the different components of ink

1. What name is given to a substance that contains atoms of different elements chemically joined together?
2. Which method of separation would you use to obtain water from a solution of copper(II) sulfate?

Atomic structure

40

Scientific models of the atom

Scientists had originally thought that atoms were tiny spheres that could not be divided.

John Dalton conducted experiments in the early 19th century and concluded that…

➤ all matter is made of indestructible atoms
➤ atoms of a particular element are identical
➤ atoms are rearranged during chemical reactions
➤ compounds are formed when two or more different types of atom join together.

Upon discovery of the electron by **J. J. Thomson** in 1897, the 'plum pudding' model suggested that the atom was a ball of positive charge with negative electrons embedded throughout.

The results from **Rutherford**, **Geiger** and **Marsden's** alpha scattering experiments (1911–1913) led to the plum pudding model being replaced by the nuclear model.

In this experiment, alpha particles (which are positive) are fired at a thin piece of gold. A few of the alpha particles do not pass through the gold and are deflected. Most went straight through the thin piece of gold. This led Rutherford, Geiger and Marsden to suggest that this is because the positive charge of the atom is confined in a small volume (now called the nucleus).

Niels Bohr adapted the nuclear model in 1913, by suggesting that electrons orbit the nucleus at specific distances. Bohr's theoretical calculations were backed up by experimental results.

Later experiments led to the idea that the positive charge of the nucleus was subdivided into smaller particles (now called protons), with each particle having the same amount of positive charge.

The work of **James Chadwick** suggested in 1932 that the nucleus also contained neutral particles that we now call neutrons.

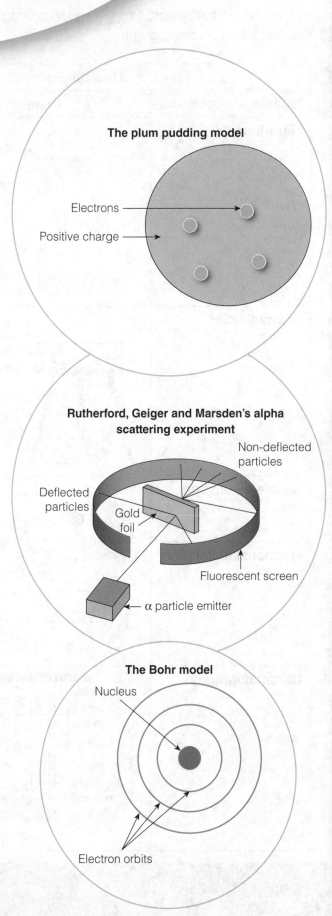

The plum pudding model

Electrons

Positive charge

Rutherford, Geiger and Marsden's alpha scattering experiment

Non-deflected particles

Deflected particles

Gold foil

Fluorescent screen

α particle emitter

The Bohr model

Nucleus

Electron orbits

Properties of atoms

Particle	Relative charge	Relative mass
Proton	+1	1
Neutron	0	1
Electron	−1	very small

➤ Atoms are neutral. This is because the number of protons is equal to the number of electrons.

➤ Atoms of different elements have different numbers of protons. This number is called the atomic number.

➤ Atoms are very small, having a radius of approximately 0.1 nm (1×10^{-10} m).

➤ The radius of the nucleus is approximately $\frac{1}{10\,000}$ of the size of the atom.

> The **mass number** tells you the total number of protons and neutrons in an atom

$$^{23}_{11}Na$$

> The **atomic number** tells you the number of protons in an atom

> **mass number – atomic number = number of neutrons**

Some atoms can have different numbers of neutrons. These atoms are called **isotopes**. The existence of isotopes results in the relative atomic mass of some elements, e.g. chlorine, not being whole numbers.

Chlorine exists as two isotopes. Chlorine-35 makes up 75% of all chlorine atoms. Chlorine-37 makes up the other 25%. We say that the abundance of chlorine-35 is 75%.

The relative atomic mass of chlorine can be calculated as follows:

$$\frac{(\text{mass of isotope 1} \times \text{abundance}) + (\text{mass of isotope 2} \times \text{abundance})}{100}$$

$$= \frac{(35 \times 75) + (37 \times 25)}{100} = 35.5$$

Keywords

Mass number ➤ The total number of protons and neutrons in an atom

Atomic number ➤ The number of protons in the nucleus of an atom

Isotopes ➤ Atoms of the same element that have the same number of protons but different numbers of neutrons

 Make models of atoms using different coloured paper to represent the protons, neutrons and electrons.

1. What is the difference between the plum pudding model of the atom and the nuclear model of the atom?

2. Why did Rutherford, Geiger and Marsden's alpha scattering experiment lead them to suggest that the positive charge in an atom was contained within a small volume?

3. How many protons, neutrons and electrons are present in the following atom?

$$^{13}_{6}C$$

4. What name is given to atoms of the same element which have the same number of protons but different numbers of neutrons?

Electronic structure & the periodic table

Keywords
Energy level ➤ A region in an atom where electrons are found
Shell ➤ Another word for an energy level

41

Electronic structures

Electrons in an atom occupy the lowest available **energy level** (**shell**). The first energy level (closest to the nucleus) can hold up to two electrons. The second and third energy levels can hold up to eight electrons.

For example, silicon has the atomic number 14. This means that there are 14 protons in the nucleus of a silicon atom and therefore there must be 14 electrons (so that the atom is neutral).

The electronic structure of silicon can be written as 2, 8, 4 or shown in a diagram like the one on the right.

Silicon is in group 4 of the periodic table. This is because it has four electrons in its outer shell. The chemical properties (reactions) of an element are related to the number of electrons in the outer shell of the atom.

The electronic structure of the first 20 elements are shown here.

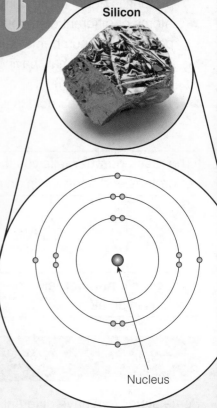

Silicon

Nucleus

Group 1	Group 2		Hydrogen, H Atomic No. = 1 No. of electrons = 1 1		Group 3	Group 4	Group 5	Group 6	Group 7	Group 8 Helium, He Atomic No. = 2 No. of electrons = 2 2
Lithium, Li Atomic No. = 3 No. of electrons = 3 2, 1	Beryllium, Be Atomic No. = 4 No. of electrons = 4 2, 2				Boron, B Atomic No. = 5 No. of electrons = 5 2, 3	Carbon, C Atomic No. = 6 No. of electrons = 6 2, 4	Nitrogen, N Atomic No. = 7 No. of electrons = 7 2, 5	Oxygen, O Atomic No. = 8 No. of electrons = 8 2, 6	Fluorine, F Atomic No. = 9 No. of electrons = 9 2, 7	Neon, Ne Atomic No. = 10 No. of electrons = 10 2, 8
Sodium, Na Atomic No. = 11 No. of electrons = 11 2, 8, 1	Magnesium, Mg Atomic No. = 12 No. of electrons = 12 2, 8, 2				Aluminium, Al Atomic No. = 13 No. of electrons = 13 2, 8, 3	Silicon, Si Atomic No. = 14 No. of electrons = 14 2, 8, 4	Phosphorus, P Atomic No. = 15 No. of electrons = 15 2, 8, 5	Sulfur, S Atomic No. = 16 No. of electrons = 16 2, 8, 6	Chlorine, Cl Atomic No. = 17 No. of electrons = 17 2, 8, 7	Argon, Ar Atomic No. = 18 No. of electrons = 18 2, 8, 8
Potassium, K Atomic No. = 19 No. of electrons = 19 2, 8, 8, 1	Calcium, Ca Atomic No. = 20 No. of electrons = 20 2, 8, 8, 2	THE TRANSITION METALS								

This table is arranged in order of atomic (proton) numbers, placing the elements in groups. Elements in the same group have the same number of electrons in their highest occupied energy level (outer shell).

The electron configuration of oxygen is 2, 6 because there are...
• 2 electrons in the first shell
• 6 electrons in the second shell.

The periodic table

The elements in the periodic table are arranged in order of increasing atomic (proton) number.
The table is called a **periodic table** because similar properties occur at regular intervals.

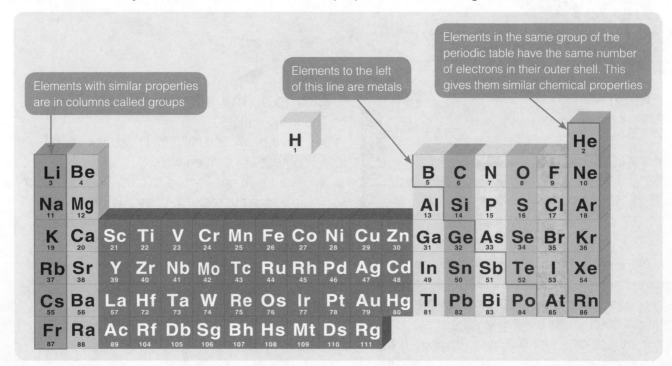

Elements with similar properties are in columns called groups

Elements to the left of this line are metals

Elements in the same group of the periodic table have the same number of electrons in their outer shell. This gives them similar chemical properties

Development of the periodic table

Before the discovery of protons, neutrons and electrons, early attempts to classify the elements involved placing them in order of their atomic weights. These early attempts resulted in incomplete tables and the placing of some elements in appropriate groups based on their chemical properties.

Dmitri Mendeleev overcame some of these problems by leaving gaps for elements that he predicted were yet to be discovered. He also changed the order for some elements based on atomic weights. Knowledge of isotopes made it possible to explain why the order based on atomic weights was not always correct.

Metals and non-metals

➤ Metals are elements that react to form positive ions.
➤ Elements that do not form positive ions are non-metals.

Typical properties of metals and non-metals	
Metals	Non-metals
Have high melting / boiling points	Have low melting / boiling points
Conduct heat and electricity	Thermal and electrical insulators
React with oxygen to form alkalis	React with oxygen to form acids
Shiny	Dull
Malleable and ductile	Brittle

Make a model of an atom of one of the first 20 elements. Make sure that you have the correct number of electrons in each shell.

1. Sodium has the atomic number 11. What is the electronic structure of sodium?
2. Why did Mendeleev leave gaps in his periodic table?
3. Element X has a high melting point, is malleable and conducts electricity. Is X a metal or a non-metal?

Groups 0, 1 and 7

42

Group 0

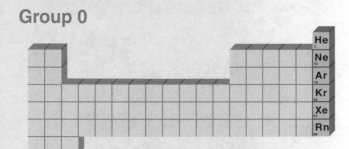

The elements in group 0 are called the **noble gases**. They are chemically inert (unreactive) and do not easily form molecules because their atoms have full outer shells (energy levels) of electrons. The inertness of the noble gases, combined with their low density and non-flammability, mean that they can be used in airships, balloons, light bulbs, lasers and advertising signs.

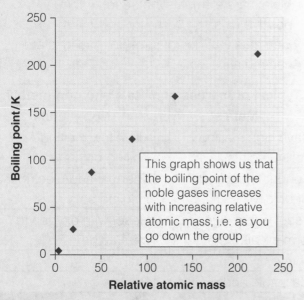

This graph shows us that the boiling point of the noble gases increases with increasing relative atomic mass, i.e. as you go down the group

Group 1 (the alkali metals)

The alkali metals…

➤ have a low density (lithium, sodium and potassium float on water)
➤ react with non-metals to form ionic compounds in which the metal ion has a charge of +1
➤ form compounds that are white solids and dissolve in water to form colourless solutions.

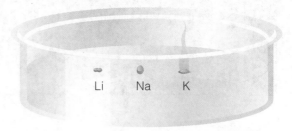

The alkali metals react with water forming metal hydroxides which dissolve in water to form alkaline solutions and hydrogen gas. For example, for the reaction between sodium and water:

$$2Na_{(s)} + 2H_2O_{(l)} \rightarrow 2NaOH_{(aq)} + H_{2(g)}$$

The alkali metals become more reactive as you go down the group because the outer shell gets further away from the positive attraction of the nucleus. This makes it easier for an atom to lose an electron from its outer shell.

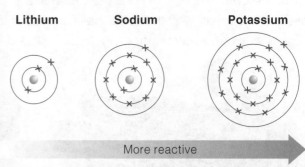

More reactive

Write each property of the alkali metals and the halogens on separate cards. Shuffle the cards. Now try to place them into the correct two piles: **alkali metals** and **halogens**.

Group 7 (the halogens)

The **halogens**…

➤ are non-metals

➤ consist of diatomic molecules (molecules made up of two atoms)

➤ react with metals to form ionic compounds where the halide ion has a charge of −1

➤ form molecular compounds with other non-metals

➤ form hydrogen halides (e.g. HCl), which dissolve in water, forming acidic solutions.

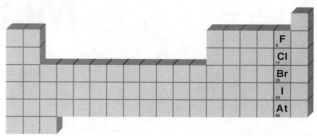

Crystals of natural fluorite

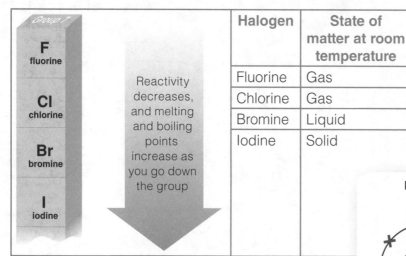

Halogen	State of matter at room temperature	Colour
Fluorine	Gas	Yellow
Chlorine	Gas	Green
Bromine	Liquid	Red / orange
Iodine	Solid	Grey / black

Halogens become less reactive as you go down the group because the outer electron shell gets further away from the attraction of the nucleus, and so an electron is gained less easily.

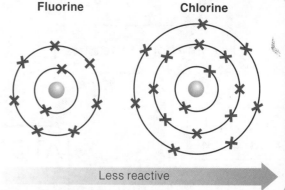

Displacement reactions of halogens

A more reactive halogen will **displace** a less reactive halogen from an aqueous solution of its metal halide.

For example:

chlorine + potassium bromide → potassium chloride + bromine
Cl_2 + 2KBr → 2KCl + Br_2

The products of reactions between halogens and aqueous solutions of halide ion salts are as follows.

		Halide salts	
	Potassium chloride, KCl	Potassium bromide, KBr	Potassium iodide, KI
Chlorine, Cl_2	No reaction	Potassium chloride + bromine	Potassium chloride + iodine
Bromine, Br_2	No reaction	No reaction	Potassium bromide + iodine
Iodine, I_2	No reaction	No reaction	No reaction

1. Why are the noble gases so unreactive?
2. What are the products of the reaction between lithium and water?
3. What is the trend in reactivity of the halogens as you go down the group?
4. What are the products of the reaction between bromine and potassium iodide?

Mind map

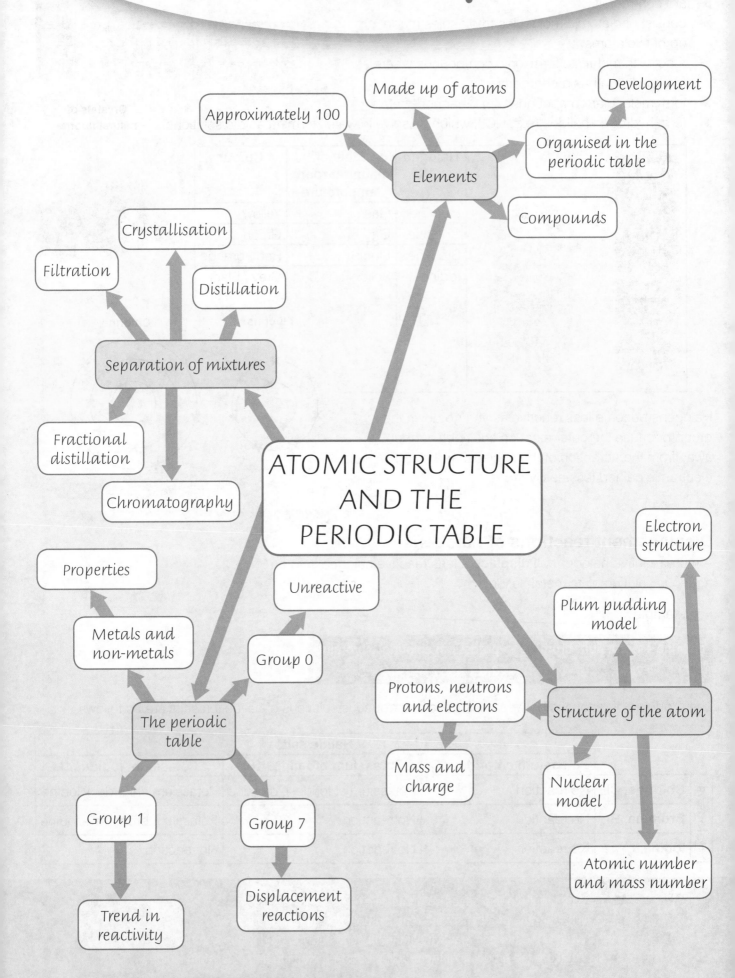

Made up of atoms

Approximately 100

Development

Organised in the periodic table

Elements

Compounds

Crystallisation

Filtration

Distillation

Separation of mixtures

Fractional distillation

Chromatography

ATOMIC STRUCTURE AND THE PERIODIC TABLE

Electron structure

Plum pudding model

Properties

Unreactive

Metals and non-metals

Group 0

Protons, neutrons and electrons

Structure of the atom

The periodic table

Mass and charge

Nuclear model

Group 1

Group 7

Atomic number and mass number

Trend in reactivity

Displacement reactions

Practice questions

1. This question is about the non-metal chlorine.

   An atom of chlorine can be represented as:

   $$^{35}_{17}Cl$$

   a) Give the number of protons, neutrons and electrons in an atom of chlorine. **(3 marks)**

   b) What is the electronic structure of chlorine? **(1 mark)**

   c) Chlorine exists as two isotopes.
   The other isotope of chlorine is represented as:

   $$^{37}_{17}Cl$$

   Give one similarity and one difference between the two isotopes of chlorine. **(2 marks)**

   d) Give one property of chlorine that is characteristic of it being a non-metal. **(1 mark)**

   e) Write a balanced symbol equation for the reaction between chlorine and sodium iodide solution. **(2 marks)**

   f) Explain why chlorine does not react with sodium fluoride solution. **(2 marks)**

2. Sand does not dissolve in water but salt does.

   a) How can sand be separated from water? **(1 mark)**

   b) Draw and label a diagram to show how water can be collected from a salt water solution. **(3 marks)**

   c) Why is crystallisation not a suitable method for obtaining water from a salt water solution? **(1 mark)**

3. The element boron exists as two naturally occurring isotopes. The mass and abundance (amount) of each isotope of boron is shown below.

Mass number of isotope	% abundance
10	20
11	80

   Use the above information to calculate the relative atomic mass of boron.
   Give your answer to three significant figures. **(3 marks)**

Chemical bonding

There are three types of chemical bond:

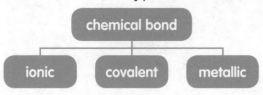

Ionic bonding

Ionic bonds occur between metals and non-metals. An ionic bond is the electrostatic force of attraction between two oppositely charged **ions** (called **cations** and **anions**).

Ionic bonds are formed when metal atoms transfer electrons to non-metal atoms. This is done so that each atom forms an ion with a full outer shell of electrons.

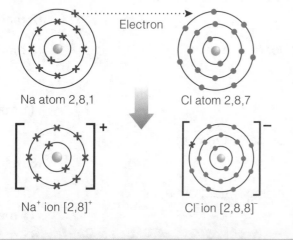

Example 1: The formation of an ionic bond between sodium and chlorine

Na atom 2,8,1

Cl atom 2,8,7

Electron

Na⁺ ion [2,8]⁺

Cl⁻ ion [2,8,8]⁻

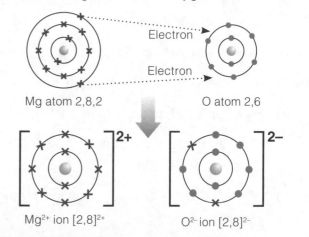

Example 2: The formation of the ionic bond between magnesium and oxygen

Mg atom 2,8,2

O atom 2,6

Electron

Electron

Mg²⁺ ion [2,8]²⁺

O²⁻ ion [2,8]²⁻

Covalent bonding

Covalent bonds occur between two non-metal atoms. Atoms share a pair of electrons so that each atom ends up with a full outer shell of electrons, such as:

➤ the formation of a covalent bond between two hydrogen atoms

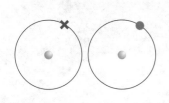

Hydrogen atoms   **A hydrogen molecule**

Covalent bond

Outermost shells overlap

➤ the covalent bonding in methane, CH_4.

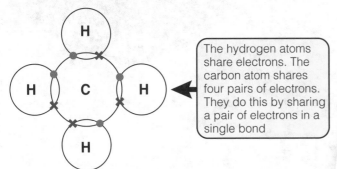

The hydrogen atoms share electrons. The carbon atom shares four pairs of electrons. They do this by sharing a pair of electrons in a single bond

Double covalent bonds occur when two pairs of electrons are shared between atoms, for example in carbon dioxide.

Carbon dioxide

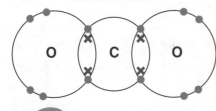

Keywords

Ion ➤ An atom or group of atoms that has gained or lost one or more electrons in order to gain a full outer shell
Cation ➤ A positive ion
Anion ➤ A negative ion
Delocalised electrons ➤ Free-moving electrons

Metallic bonding

Metals consist of giant structures. Each atom loses its outer shell electrons and these electrons become **delocalised**, i.e. they are free to move through the structure. The metal cations are arranged in a regular pattern called a lattice (see below).

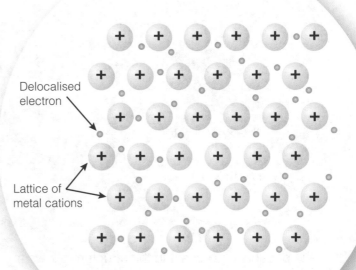

Delocalised electron

Lattice of metal cations

Use the models of atoms you have previously made to try to work out what happens when different atoms bond together. Make sure that each atom ends up with a full outer shell. Remember that you might need more than one of each type of atom.

1. What type of bonding occurs between a metal and non-metal atom?
2. How many electrons are present in every covalent bond?
3. Describe the structure of a metal.

Ionic and covalent structures

44

Structure of ionic compounds

Compounds containing ionic bonds form giant structures. These are held together by strong electrostatic forces of attraction between the oppositely charged ions. These forces act in all directions throughout the lattice.

The diagrams below represent a typical giant ionic structure, sodium chloride.

Sodium chloride

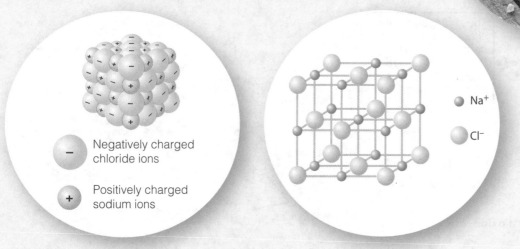

Negatively charged chloride ions

Positively charged sodium ions

Na^+

Cl^-

The ratio of each ion present in the structure allows the **empirical formula** of the compound to be worked out. In the diagrams above, for every Na^+ ion there is a Cl^- ion. This means that the empirical formula is NaCl.

Structure of covalent compounds

Covalently bonded substances may consist of….

➤ small molecules / simple molecular structures (e.g. Cl_2, H_2O and CH_4)
➤ large molecules, called **polymers**
➤ giant covalent structures (e.g. diamond, graphite and silicon dioxide).

Small molecular structures

The bonding between hydrogen and carbon in methane can be represented in several ways, as shown here.

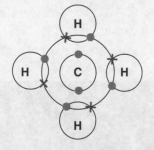 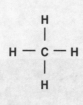 

Large molecules

Polymers can be represented in the form:

where n is a large number

V, W, X and Y represent the atoms bonded to the carbon atoms

For example, poly(ethene) can be represented as:

Giant covalent structures

This is the giant covalent structure of silicon dioxide.

Keywords

Empirical formula ➤ The simplest whole number ratio of each kind of atom present in a compound

Polymer ➤ A large, long-chained molecule

Make a model that represents sodium chloride or silicon dioxide. Count the number of each particle and then work out the simplest whole number ratio of each particle present. This is the empirical formula.

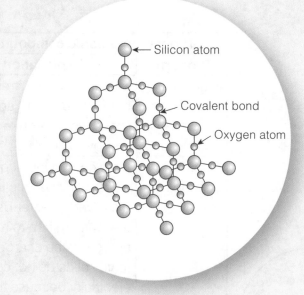

← Silicon atom

← Covalent bond

← Oxygen atom

Diamond

The formula of silicon dioxide is SiO_2 – this can be deduced by looking at the ratio of Si to O atoms in the diagram above.

Models, such as dot-and-cross diagrams, ball-and-stick diagrams and two- / three-dimensional diagrams to represent structures, are limited in value, as they do not accurately represent the structures of materials. For example, a chemical bond is not a solid object as depicted in some models, it is actually an attraction between particles. The relative size of different atoms is often not shown when drawing diagrams.

Graphite

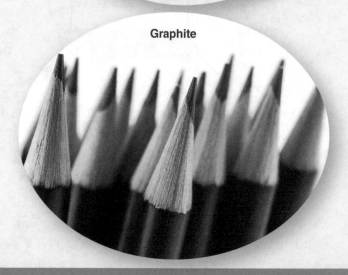

1. What forces hold ionic structures together?
2. What are the three types of covalent substance?
3. Draw a diagram to show the structure of sodium chloride.

States of matter: properties of compounds

States of matter

The three main states of matter are **solids**, **liquids** and **gases**. Individual atoms do not have the same properties as these bulk substances. The diagram below shows how they can be interconverted and also how the particles in the different states of matter are arranged.

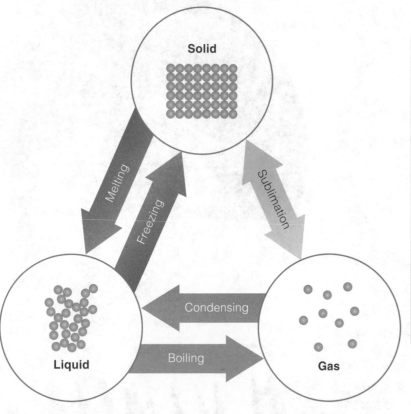

Properties of ionic compounds	
Property	**Explanation**
High melting and boiling points	There are lots of strong bonds throughout an ionic lattice which require lots of energy to break.
Electrical conductivity	Ionic compounds conduct electricity when molten or dissolved in water because the ions are free to move and carry the charge. Ionic solids do not conduct electricity because the ions are in a fixed position and are unable to move.

These are physical changes because the particles are either gaining or losing energy and are not undergoing a chemical reaction. Particles in a gas have more energy than in a liquid; particles in a liquid would have more energy than in a solid.

Melting and freezing occur at the same temperature. Condensing and boiling also occur at the same temperature. The amount of energy needed to change state depends on the strength of the forces between the particles of the substance.

The stronger the forces between the particles, the higher the melting and boiling points of the substance.

Keyword
Intermolecular forces ➤ The weak forces of attraction that occur between molecules

Properties of small molecules

Substances made up of small molecules are usually gases or liquids at room temperature. They have relatively low melting and boiling points because there are weak (intermolecular) forces that act between the molecules. It is these weak forces and not the strong covalent bonds that are broken when the substance melts or boils.

Substances made up of small molecules do not normally conduct electricity. This is because the molecules do not have an overall electric charge or delocalised electrons.

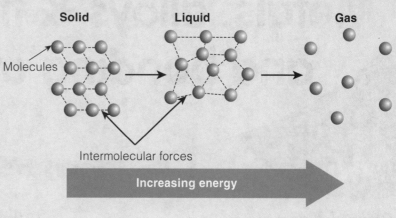

Solid   **Liquid**   **Gas**

Molecules

Intermolecular forces

Increasing energy

Polymers

Polymers are very large molecules made up of atoms joined together by strong covalent bonds. The **intermolecular forces** between polymer molecules are much stronger than in small molecules because the molecules are larger. This is why most polymers are solid at room temperature.

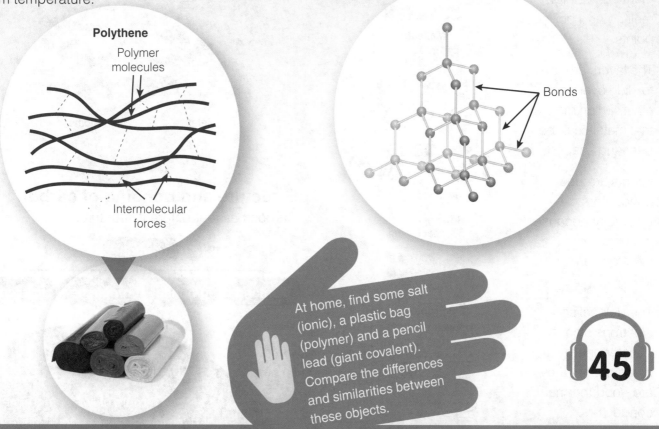

Polythene

Polymer molecules

Intermolecular forces

Giant covalent structures

Substances with a giant covalent structure are solids at room temperature. They have relatively high melting and boiling points. This is because there are lots of strong covalent bonds that need to be broken.

Bonds

At home, find some salt (ionic), a plastic bag (polymer) and a pencil lead (giant covalent). Compare the differences and similarities between these objects.

45

1. What is the main factor that determines the melting point of a solid?
2. Why do ionic compounds have relatively high melting points?
3. What needs to be broken in order to melt a substance made up of small molecules such as water?
4. Why do polymers have higher melting points than substances made up of small molecules?
5. Why do substances with giant covalent structures have relatively high melting points?

Metals, alloys & the structure and bonding of carbon

46

Structure and properties of metals

Metals have giant structures. Metallic bonding (the attraction between the cations and the delocalised electrons) is strong meaning that most metals have high melting and boiling points.

The layers are able to slide over each other, which means that metals can be bent and shaped.

Metals are good conductors of electricity because the delocalised electrons are able to move.

The delocalised electrons also transfer energy meaning that they are good thermal conductors.

High melting point

Strong forces of attraction between cations and electrons

Malleable

Force applied

Force applied

Rows of ions slide over each other

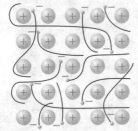

Electrons moving between the cations

Alloys

Most metals we use are **alloys**. Many pure metals (such as gold, iron and aluminium) are too soft for many uses and so are mixed with other materials (usually metals) to make alloys.

The different sizes of atoms in alloys make it difficult for the layers to slide over each other. This is why alloys are harder than pure metals.

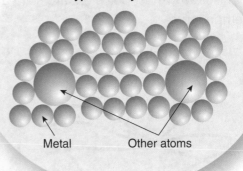

Typical alloy structure

Metal   Other atoms

Structure and bonding of carbon

Carbon has four different structures.

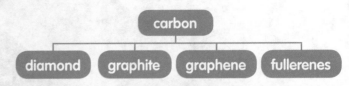

carbon

diamond   graphite   graphene   fullerenes

Keywords

Alloy ➤ A mixture of two or more metals, or a mixture of a metal and a non-metal

Fullerene ➤ A molecule made of carbon atoms arranged as a hollow sphere

Nanotube ➤ A molecule made of carbon atoms arranged in a tubular structure

High tensile strength ➤ Does not break easily when stretched

Diamond

Diamond is a giant covalent structure (or macromolecule) where each carbon atom is bonded to four others.

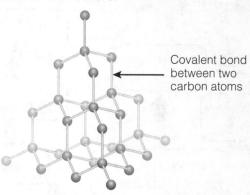

Covalent bond between two carbon atoms

In diamond, there are lots of very strong covalent bonds so diamond...

➤ is hard
➤ has a high melting point.

For these reasons, diamond is used in making cutting tools.

There are no free electrons in diamond so it does not conduct electricity.

Graphite

Graphite is also a giant covalent structure, with each carbon atom forming three covalent bonds, resulting in layers of hexagonal rings of carbon atoms. Carbon has four electrons in its outer shell and as only three are used for bonding the other one is delocalised.

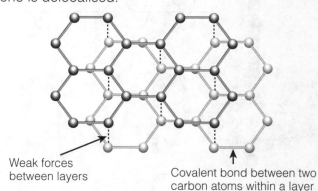

Weak forces between layers

Covalent bond between two carbon atoms within a layer

The layers in graphite are able to slide over each other because there are only weak intermolecular forces holding them together. This is why graphite is soft and slippery. These properties make graphite suitable for use as a lubricant.

Like diamond, there are lots of strong covalent bonds in graphite so it has a high melting point.

The delocalised electrons allow graphite to conduct electricity and heat.

Graphene and fullerenes

Graphene is a single layer of graphite and so it is one atom thick.

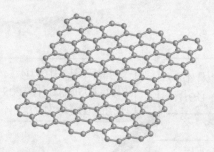

Fullerenes are molecules made up of carbon atoms and they have hollow shapes. The structure of fullerenes is based on hexagonal rings of carbon atoms but they may also contain rings with five or seven carbon atoms.

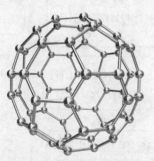

buckminsterfullerene (C_{60}) was the first fullerene to be discovered

Carbon **nanotubes** are cylindrical fullerenes.

Fullerenes have **high**...

➤ **tensile strength**
➤ electrical conductivity
➤ thermal conductivity.

Fullerenes can be used...

➤ for drug delivery into the body
➤ as lubricants
➤ for reinforcing materials, e.g. tennis rackets.

Make flash cards with the four different structures of carbon on one side and the properties / uses on the reverse. Use these cards to test your knowledge.

1. Why do metals generally have high melting points?
2. Why are alloys harder than pure metals?
3. Why does graphite conduct electricity?
4. State two properties of fullerenes.
5. Give two uses of fullerenes.

Mind map

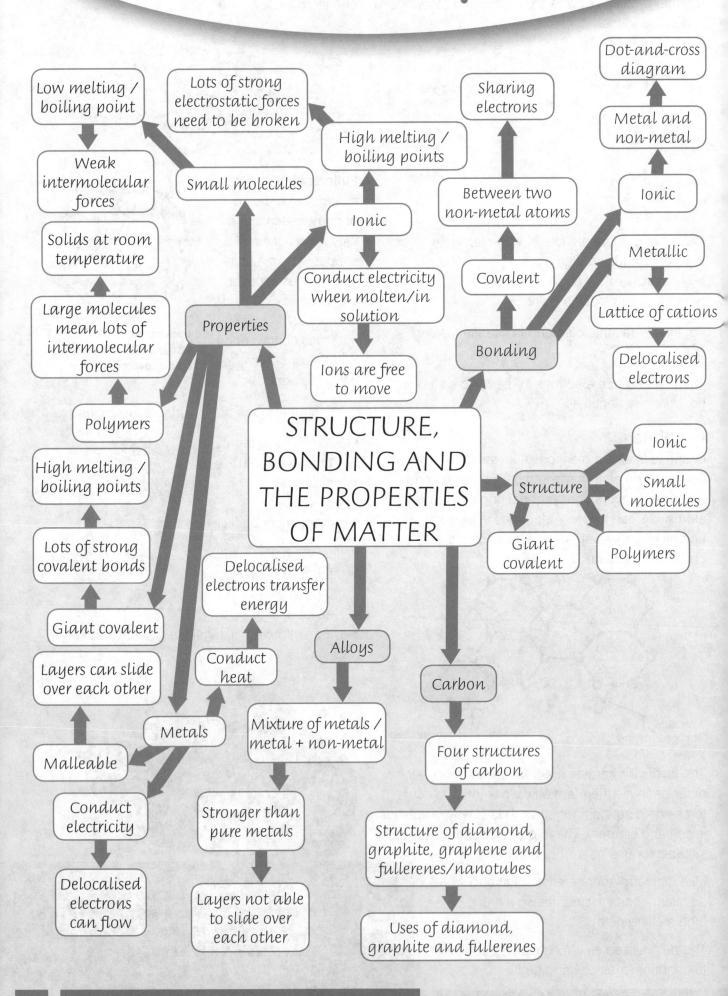

Low melting / boiling point

Lots of strong electrostatic forces need to be broken

Weak intermolecular forces

Small molecules

Sharing electrons

Dot-and-cross diagram

Metal and non-metal

High melting / boiling points

Solids at room temperature

Ionic

Between two non-metal atoms

Ionic

Metallic

Large molecules mean lots of intermolecular forces

Properties

Conduct electricity when molten/in solution

Covalent

Bonding

Lattice of cations

Polymers

Ions are free to move

Delocalised electrons

High melting / boiling points

STRUCTURE, BONDING AND THE PROPERTIES OF MATTER

Structure

Ionic

Small molecules

Lots of strong covalent bonds

Delocalised electrons transfer energy

Giant covalent

Polymers

Giant covalent

Layers can slide over each other

Conduct heat

Alloys

Carbon

Malleable

Metals

Mixture of metals / metal + non-metal

Four structures of carbon

Conduct electricity

Stronger than pure metals

Structure of diamond, graphite, graphene and fullerenes/nanotubes

Delocalised electrons can flow

Layers not able to slide over each other

Uses of diamond, graphite and fullerenes

Practice questions

1. This question is about the structure and bonding of sodium chloride and the properties that it has.

    a) What type of bonding is present in sodium chloride? **(1 mark)**

    b) Draw a dot-and-cross diagram to show the formation of the bond in sodium chloride. Draw the electronic structure of the atoms before the bond has formed and the ions after the bond has formed. **(3 marks)**

    c) Explain how the ions in sodium chloride are held together. **(2 marks)**

    d) Does sodium chloride have a high or a low boiling point? Explain your answer. **(2 marks)**

    e) Draw a diagram showing the positions of the ions in a crystal of sodium chloride. **(1 mark)**

2. Carbon dioxide and silicon dioxide have different structures.

    a) What type of bonding is present in both carbon dioxide and silicon dioxide? **(1 mark)**

    b) The diagram below shows the bonding in carbon dioxide.

    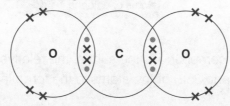

    i)   How many covalent bonds are there between the carbon atom and each oxygen atom? **(1 mark)**

    ii) Does carbon dioxide have a simple molecular or giant covalent (macromolecular) structure? **(1 mark)**

    c) The structure of silicon dioxide is shown below.

    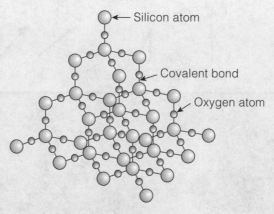

    i)   Does silicon dioxide have a simple molecular or a giant covalent (macromolecular) structure? **(1 mark)**

    ii) In terms of structure, what is the difference between a simple molecular structure and a giant covalent (macromolecular) structure? **(2 marks)**

    d) Would you expect silicon dioxide to have a higher or lower boiling point than carbon dioxide? Explain your answer. **(2 marks)**

Mass and equations

47

Keywords

Relative formula mass ➤ The sum of the atomic masses of the atoms in a formula

Thermal decomposition ➤ The breakdown of a chemical substance due to the action of heat

Conservation of mass

The total mass of reactants in a chemical reaction is equal to the total mass of the products because atoms are not created or destroyed.

Chemical reactions are represented by balanced symbol equations.

For example:

This means there are four atoms of Na ➤ **$4Na + TiCl_4 \rightarrow 4NaCl + Ti$**

This means there are four atoms of chlorine in $TiCl_4$

Reactants   Products

Relative formula mass

The **relative formula mass** (M_r) of a compound is the sum of the relative atomic masses (see the periodic table on page 232) of the atoms in the formula.

For example:
➤ The M_r of MgO is 40   (24 + 16)
➤ The M_r of H_2SO_4 is 98 $[(2 \times 1) + 32 + (4 \times 16)]$

In a balanced symbol equation, the sum of the relative formula masses of the reactants equals the sum of the relative formula masses of the products.

Example: $CaCO_3 + 2HCl \rightarrow CaCl_2 + H_2O + CO_2$

Sum of relative formula masses: $100 + (2 \times 36.5) = 173$ $111 + 18 + 44 = 173$

Mass changes when a reactant or product is a gas

During some chemical reactions, there can appear to be a change in mass. When copper is heated its mass actually increases because oxygen is being added to it.

Copper

> **copper + oxygen → copper oxide**

The mass of copper oxide formed is equal to the starting mass of copper plus the mass of the oxygen that has been added to it.

During a **thermal decomposition** reaction of a metal carbonate, the final mass of remaining metal oxide solid is less than the starting mass. This is because when the metal carbonate thermally decomposes it releases carbon dioxide gas into the atmosphere.

> **copper carbonate → copper oxide + carbon dioxide**

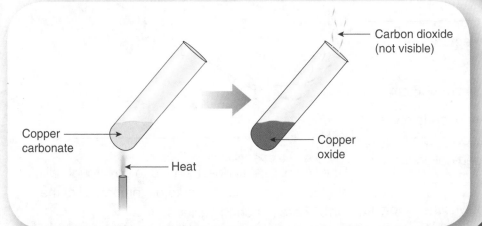

Carbon dioxide (not visible)

Copper carbonate

Heat

Copper oxide

Make cards with the symbol of the element on one side and the relative atomic mass on the reverse.
Use these cards to work out the relative formula mass of the compounds you find in this book.

Example:

Starting mass of copper carbonate   = 8.00 g
Final mass of copper oxide   = 5.15 g

Therefore, mass of carbon dioxide
released to the atmosphere   = 2.85 g (8.00 g – 5.15 g)

---

1. In a chemical reaction the total mass of reactants is 13.60 g. Will the expected mass of all the products be lower, higher or the same as 13.60 g?
2. With reference to the periodic table on page 232 work out the relative formula mass of the following compounds.
   a) NH_4NO_3
   b) $Mg(OH)_2$
3. When 5.0 g of zinc carbonate is heated will the mass of remaining metal oxide be lower, higher or the same as 5.0 g?

Chemical measurements & Concentration of solutions

48

Chemical measurements

When a measurement is made during a chemical experiment there is always some uncertainty about the result obtained.

For example, in an experiment to investigate the volume of hydrogen produced when a 2 cm strip of magnesium was reacted with 10 cm^3 of acid, the following results were obtained:

Experiment	1	2	3	4
Volume of hydrogen collected / cm^3	29	28	33	34

The **mean** volume of hydrogen collected is (29 + 28 + 33 + 34) ÷ 4 = 31 cm^3

The **range** of results is 28 cm^3 to 34 cm^3 = 6 cm^3

Therefore, a reasonable estimate of uncertainty is half the range of results. In this case we can say that the average volume of hydrogen collected was 31 cm^3 ± 3 cm^3.

The level of uncertainty can be shown when plotting results on a graph.

The graph below shows experimental results and how the uncertainty can be indicated by drawing **error bars.**

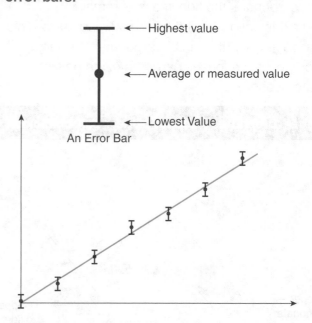

An Error Bar

Highest value
Average or measured value
Lowest Value

The line of best fit doesn't need to go through all of the data points, but it should go through all of the **error bars** (ignoring outliers).

There are also errors associated with any measuring instrument, e.g. a balance. The uncertainty is usually half the smallest scale division.

A balance reads to 0.01g.
The uncertainty is therefore ± 0.005g.

Keywords

Mean ➤ The sum of all the values divided by the number of values
Range ➤ The difference between the largest and smallest values
Error bars ➤ The upper limit is the mean plus half the range. The lower limit is the mean minus half the range
Solute ➤ A solid that has dissolved in a liquid to form a solution

Concentration of solutions in g/dm³

Many chemical reactions take place in solutions. The concentration of a solution can be measured in mass of **solute** per given volume of solution, e.g. grams per dm³ (1 dm³ = 1000 cm³).

For example, a solution of 5 g/dm³ has 5 g of solute dissolved in 1 dm³ of water. It has half the concentration of a 10 g/dm³ solution of the same solute.

The mass of solute in a solution can be calculated if the concentration and volume of solution are known.

> **Example:** Calculate the mass of solute in 250 cm³ of a solution whose concentration is 8 g/dm³.
>
> **Step 1:** Divide the mass by 1000 (this gives you the mass of solute in 1 cm³).
>
> $$8 \div 1000 = 0.008 \text{ g/cm}^3$$
>
> **Step 2:** Multiply this value by the volume specified.
>
> $$0.008 \times 250 = 2 \text{ g}$$

Look at apparatus in your school laboratory that is used to measure amounts of chemicals and see if you can determine the uncertainty in their measurements.

1. A burette reads to 0.10 cm³. What is the level of uncertainty in this measurement?
2. An experiment was carried out four times. The reading on the burette in each experiment was 22.40 cm³, 20.80 cm³, 22.60 cm³ and 20.20 cm³. What is the average burette reading and the range of results?

Mind map

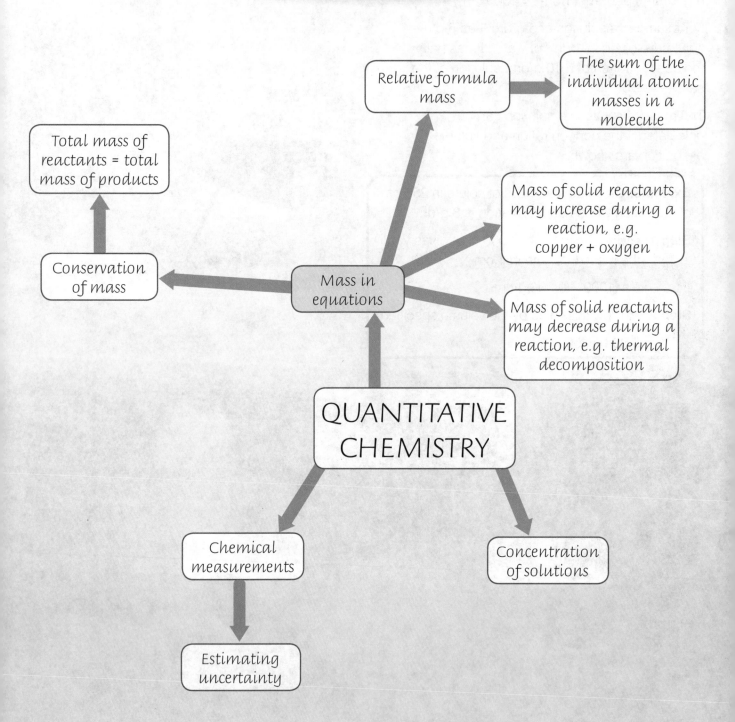

Relative formula mass → The sum of the individual atomic masses in a molecule

Total mass of reactants = total mass of products

Conservation of mass

Mass in equations

Mass of solid reactants may increase during a reaction, e.g. copper + oxygen

Mass of solid reactants may decrease during a reaction, e.g. thermal decomposition

QUANTITATIVE CHEMISTRY

Chemical measurements

Estimating uncertainty

Concentration of solutions

Practice questions

1. Calcium carbonate thermally decomposes when heated to form calcium oxide and carbon dioxide gas.

   The equation for the reaction is shown below.

   $$CaCO_{3(s)} \rightarrow CaO_{(s)} + CO_{2(g)}$$

   In an experiment, a student placed 10 g of $CaCO_3$ in a crucible and heated it strongly for 10 minutes.

   a) Will the mass of solid in the crucible at the end of the experiment be higher or lower than 10 g? Explain your answer. **(2 marks)**

   b) With reference to the periodic table (see page 232), what is the relative formula mass of calcium carbonate? **(1 mark)**

   c) Calculate the mass of carbon dioxide that would be produced in this experiment. **(2 marks)**

2. A student placed a piece of copper metal weighing 1.20g into a Bunsen flame. After 2 minutes she stopped heating, allowed the copper to cool and reweighed it. It now weighed 1.25g.

   a) Explain why the mass of copper increased. **(1 mark)**

   b) By how much did its mass increase? **(1 mark)**

   c) Write a word equation for the reaction that takes place in this experiment. **(2 marks)**

   d) How does the law of conservation of mass explain the increase of mass in this experiment? **(1 mark)**

3. 75 cm³ of a solution contains 21g of dissolved solute.

   a) Calculate the mass of solute that is present in 100 cm³ of the same solution. **(1 mark)**

   b) What is the concentration of the solution in g/dm³? **(1 mark)**

Reactivity of metals and metal extraction

49

Keywords

Oxidation ➤ A reaction involving the gain of oxygen

Reduction ➤ A reaction involving the loss of oxygen

Reaction of metals with oxygen

Many metals react with oxygen to form metal oxides, for example:

> **copper + oxygen ⟶ copper oxide**

These reactions are called **oxidation** reactions. Oxidation reactions take place when a chemical gains oxygen.

When a substance loses oxygen it is called a **reduction** reaction.

The reactivity series

When metals react they form positive ions. The more easily the metal forms a positive ion the more reactive the metal.

Calcium and magnesium are both in group 2 of the periodic table so will form 2⁺ ions when they react. Calcium is more reactive than magnesium so it has a greater tendency/is more likely to form the 2⁺ ion.

Decreasing reactivity →

Metal	Reaction with water	Reaction with acid
Potassium	Very vigorous	Explosive
Sodium	Vigorous	Dangerous
Lithium	Steady	Very vigorous
Calcium	Steady fizzing and bubbling	Vigorous
Magnesium	Slow reaction	Steady fizzing and bubbling
*Carbon		
Zinc	Very slow reaction	Gentle fizzing and bubbling
Iron	Extremely slow	Slight fizzing and bubbling
*Hydrogen		
Copper	No reaction	No reaction
Gold	No reaction	No reaction

* included for comparison

The reactivity series can also be used to predict displacement reactions.

> **zinc + copper oxide ⟶ zinc oxide + copper**

In this reaction…
➤ zinc displaces (i.e. takes the place of) copper
➤ zinc is oxidised (i.e. it gains oxygen)
➤ copper oxide is reduced (i.e. it loses oxygen).

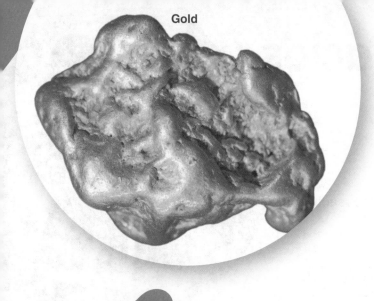

Gold

Extraction of metals and reduction

Unreactive metals, such as gold, are found in the Earth's crust as pure metals. Most metals are found as compounds and chemical reactions are required to extract the metal. The method of extraction depends on the position of the metal in the reactivity series.

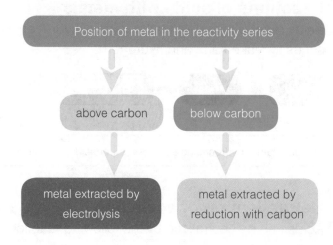

Position of metal in the reactivity series

above carbon | below carbon

metal extracted by electrolysis | metal extracted by reduction with carbon

For example, iron is found in the earth as iron(III) oxide, Fe_2O_3. The iron(III) oxide can be reduced by reacting it with carbon.

Pupils Sometimes Like Chasing Mischievous Zebras In Country Gardens

Make up your own mnemonic for the reactivity series!

The Fe_2O_3 is reduced to Fe

$$2Fe_2O_3 + 3C \rightarrow 4Fe + 3CO_2$$

The C is oxidised to CO_2

1. Write a word equation for the oxidation of magnesium to form magnesium oxide.
2. Both sodium and lithium react to form 1+ ions. Which one of these metals has a greater tendency / is more likely to form this ion?
3. Name two metals from the reactivity series that are extracted by reduction with carbon.

Reactions of acids

Reactions of acids with metals

Acids react with metals that are above hydrogen in the reactivity series to make **salts** and hydrogen.

Acid ➕ Metal ➡ Salt + Hydrogen

This is a salt

For example:

magnesium + hydrochloric acid ⟶ magnesium chloride + hydrogen

Neutralisation of acids and the preparation of soluble salts

Acids can be neutralised by the following reactions.

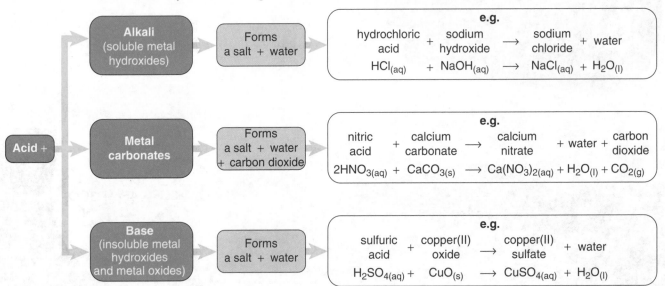

Acid +

Alkali (soluble metal hydroxides) ➡ Forms a salt + water ➡

e.g.
hydrochloric acid + sodium hydroxide ⟶ sodium chloride + water
$HCl_{(aq)}$ + $NaOH_{(aq)}$ ⟶ $NaCl_{(aq)}$ + $H_2O_{(l)}$

Metal carbonates ➡ Forms a salt + water + carbon dioxide ➡

e.g.
nitric acid + calcium carbonate ⟶ calcium nitrate + water + carbon dioxide
$2HNO_{3(aq)}$ + $CaCO_{3(s)}$ ⟶ $Ca(NO_3)_{2(aq)}$ + $H_2O_{(l)}$ + $CO_{2(g)}$

Base (insoluble metal hydroxides and metal oxides) ➡ Forms a salt + water ➡

e.g.
sulfuric acid + copper(II) oxide ⟶ copper(II) sulfate + water
$H_2SO_{4(aq)}$ + $CuO_{(s)}$ ⟶ $CuSO_{4(aq)}$ + $H_2O_{(l)}$

The first part of the salt formed contains the positive ion (usually the metal) from the alkali, base or carbonate followed by…

➤ chloride if hydrochloric acid was used

➤ sulfate if sulfuric acid was used

➤ nitrate if nitric acid was used.

For example, when calcium hydroxide is reacted with sulfuric acid, the salt formed is calcium sulfate.

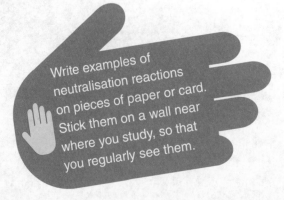

Write examples of neutralisation reactions on pieces of paper or card. Stick them on a wall near where you study, so that you regularly see them.

Preparation of soluble salts

Soluble salts can be prepared by the following method.

> Add solid to the acid until no more reacts ➤ Filter off the excess solid ➤ Obtain the solid salt by **crystallisation**

For example, copper(II) sulfate crystals can be made by reacting copper(II) oxide with sulfuric acid.

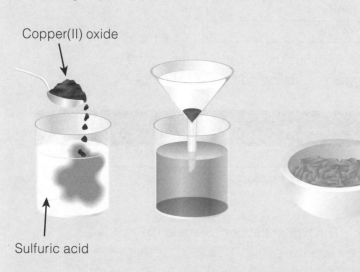

Copper(II) oxide

Sulfuric acid

> Add copper(II) oxide to sulfuric acid ➤ Filter to remove any unreacted copper(II) oxide ➤ Evaporate to leave behind blue crystals of the 'salt' copper(II) sulfate

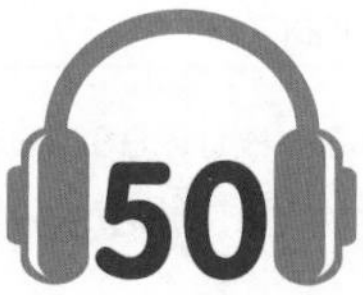

Copper(II) sulfate

50

1. Write a word equation for the reaction that occurs when zinc reacts with sulfuric acid.
2. Other than an alkali, name a substance that can neutralise an acid.
3. Name the salt formed when lithium oxide reacts with nitric acid.
4. What are the three main steps in preparing a soluble salt from an acid and a metal oxide?

pH, neutralisation, acid strength and electrolysis

Indicators, the pH scale and neutralisation reactions

Indicators are useful dyes that become different colours in acids and alkalis.

Indicator	Colour in acid	Colour in alkali
Litmus	Red	Blue
Phenolphthalein	Colourless	Pink
Methyl orange	Pink	Yellow

The pH scale measures the acidity or alkalinity of a solution. The pH scale runs from 0 to 14 and the pH of a solution can be measured using universal indicator or a pH probe.

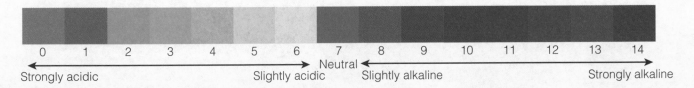

Acids are solutions that contain hydrogen ions (H^+). The higher the concentration of hydrogen ions, the more acidic the solution (i.e. the lower the pH).

Alkalis are solutions that contain hydroxide ions (OH^-). The higher the concentration of hydroxide ions, the more alkaline the solution (i.e. the higher the pH).

When an acid is neutralised by an alkali, the hydrogen ions from the acid react with the hydroxide ions in the alkali to form water.

$$H^+_{(aq)} + OH^-_{(aq)} \rightarrow H_2O_{(l)}$$

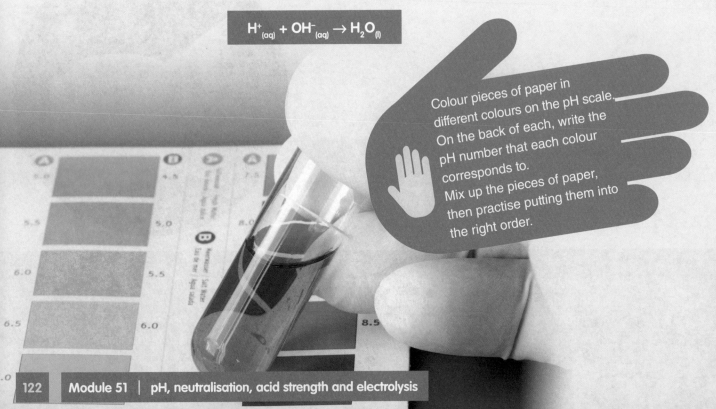

Colour pieces of paper in different colours on the pH scale. On the back of each, write the pH number that each colour corresponds to.
Mix up the pieces of paper, then practise putting them into the right order.

Electrolysis

An electric current is the flow of electrons through a conductor but it can also flow by the movement of ions through a solution or a liquid.

Covalent compounds do not contain free electrons or ions that can move. So they will not conduct electricity when solid, liquid, gas or in solution.

The ions in:
➤ an **ionic solid** are fixed and cannot move
➤ an **ionic substance** that is **molten** or in **solution** are free to move.

Electrolysis is a chemical reaction that involves passing electricity through an **electrolyte**. An electrolyte is a liquid that conducts electricity. Electrolytes are either molten ionic compounds or solutions of ionic compounds. Electrolytes are decomposed during electrolysis.
➤ The positive ions (**cations**) move to, and discharge at, the negative electrode (**cathode**).
➤ The negative ions (**anions**) move to, and discharge at, the positive electrode (**anode**).

Electrons are removed from the anions at the anode. These electrons then flow around the circuit to the cathode and are transferred to the cations.

Keywords

Electrolyte ➤ A liquid or solution containing ions that is broken down during electrolysis

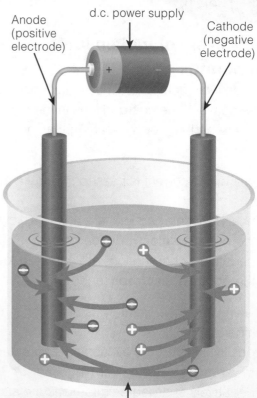

Anode (positive electrode)

d.c. power supply

Cathode (negative electrode)

Electrolyte (liquid that conducts electricity and decomposes in electrolysis)

1. What type of substance has a pH of less than 7?
2. Which ion is responsible for solutions being alkaline?
3. Write an ionic equation for the reaction that takes place when an acid is neutralised by an alkali.
4. Why must electrolytes be in the liquid state?

Applications of electrolysis

52

Keywords

Cathode ➤ The negative electrode
Anode ➤ The positive electrode

Electrolysis of molten ionic compounds

When an ionic compound melts, electrostatic forces between the charged ions in the crystal lattice are broken down, meaning that the ions are free to move.

When a direct current is passed through a molten ionic compound:

➤ positively charged ions are attracted towards the **negative electrode** (cathode)

➤ negatively charged ions are attracted towards the **positive electrode** (anode).

For example, in the electrolysis of molten lead bromide:

➤ positively charged lead ions are attracted towards the **cathode**, forming lead

➤ negatively charged bromide ions are attracted towards the **anode**, forming bromine.

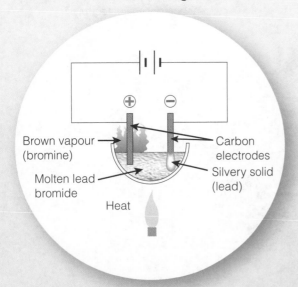

Brown vapour (bromine)
Carbon electrodes
Molten lead bromide
Silvery solid (lead)
Heat

When ions get to the oppositely charged electrode they are **discharged** – they lose their charge. For example, in the electrolysis of molten lead bromide, the non-metal ion loses electrons to the positive electrode to form a bromine atom. The bromine atom then bonds with a second atom to form a bromine molecule.

(ws) Using electrolysis to extract metals

Aluminium is the most abundant metal in the Earth's crust. It must be obtained from its ore by electrolysis because it is too reactive to be extracted by heating with carbon. The electrodes are made of graphite (a type of carbon). The aluminium ore (bauxite) is purified to leave aluminium oxide, which is then melted so that the ions can move. Cryolite is added to increase the conductivity and lower the melting point.

When a current passes through the molten mixture:

➤ positively charged aluminium ions move towards the negative electrode (**cathode**) and form aluminium

➤ negatively charged oxygen ions move towards the positive electrode (**anode**) and form oxygen.

The positive electrodes gradually wear away (because the graphite electrodes react with the oxygen to form carbon dioxide gas). This means they have to be replaced every so often. Extracting aluminium can be quite an expensive process because of the cost of the large amounts of electrical energy needed to carry it out.

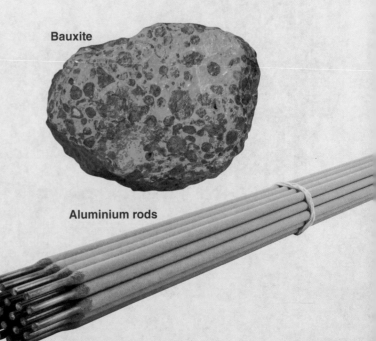

Bauxite

Aluminium rods

Electrolysis of aqueous solutions

When a solution undergoes electrolysis, there is also water present. During electrolysis water molecules break down into hydrogen ions and hydroxide ions.

$$H_2O_{(l)} \rightarrow H^+_{(aq)} + OH^-_{(aq)}$$

This means that when an aqueous compound is electrolysed there are two cations present (H^+ from water and the metal cation from the compound) and two anions present (OH^- from water and the anion from the compound).

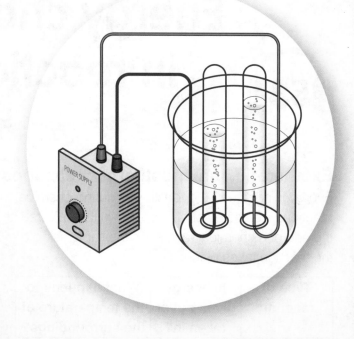

Solution	Product at cathode	Product at anode
copper chloride	copper	chlorine
sodium sulfate	hydrogen	oxygen
water (diluted with sulfuric acid to aid conductivity)	hydrogen	oxygen

For example, in copper(II) sulfate solution…
➤ cations present: Cu^{2+} and H^+
➤ anions present: SO_4^{2-} and OH^-

At the positive electrode (anode): *Oxygen is produced unless the solution contains halide ions.* In this case the oxygen is produced.

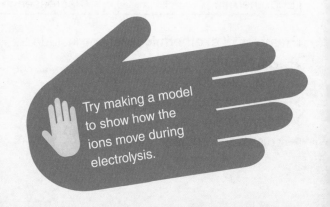

Try making a model to show how the ions move during electrolysis.

At the negative electrode (cathode): *The least reactive element is formed.* The reactivity series in Module 49 will be helpful here. In this case, hydrogen is formed.

Example:

What are the three products of the electrolysis of sodium chloride solution?

Cations present: Na^+ and H^+
Q. What happens at the cathode?
A. Hydrogen is less reactive than sodium therefore hydrogen gas is formed.

Anions present: Cl^- and OH^-
Q. What happens at the anode?
A. A halide ion (Cl^-) is present therefore chlorine will be formed.

The sodium ions and hydroxide ions stay in solution (i.e. sodium hydroxide solution remains).

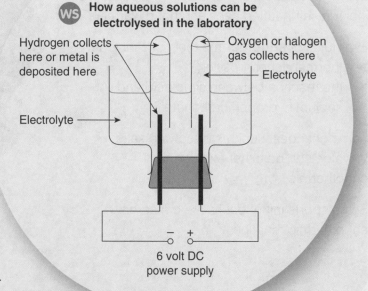

WS How aqueous solutions can be electrolysed in the laboratory

Hydrogen collects here or metal is deposited here

Oxygen or halogen gas collects here

Electrolyte

Electrolyte

6 volt DC power supply

---

1. When molten potassium iodide is electrolysed, what will be formed at the cathode and anode?
2. When aqueous potassium iodide is electrolysed, what will be formed at the cathode and anode?

Energy changes in reactions

53

Keywords

Exothermic ➤ A reaction in which energy is given out
Endothermic ➤ A reaction in which energy is taken in

Exothermic and endothermic reactions

Energy is not created or destroyed during chemical reactions, i.e. the amount of energy in the universe at the end of a chemical reaction is the same as before the reaction takes place.

Type of reaction	Is energy given out or taken in?	What happens to the temperature of the surroundings?
Exothermic	out	increases
Endothermic	in	decreases

Examples of **exothermic** reactions include…
➤ combustion
➤ neutralisation
➤ many oxidation reactions
➤ precipitation reactions
➤ displacement reactions.

Everyday applications of exothermic reactions include self-heating cans and hand warmers.

Examples of **endothermic** reactions include…
➤ thermal decomposition
➤ the reaction between citric acid and sodium hydrogen carbonate.

Some changes, such as dissolving salts in water, can be either exothermic or endothermic.

Some sports injury packs are based on endothermic reactions.

Reaction profiles

For a chemical reaction to occur, the reacting particles must collide together with sufficient energy. The minimum amount of energy that the particles must have in order to react is known as the 'activation energy'.

Reaction profiles can be used to show the relative energies of reactants and products, the activation energy and the overall energy change of a reaction.

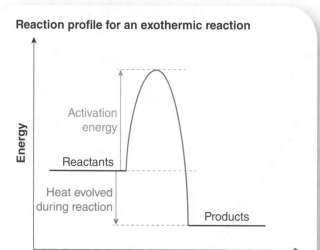

Reaction profile for an exothermic reaction

Energy / Progress of reaction — Activation energy, Reactants, Heat evolved during reaction, Products

Chemical reactions in which more energy is made when new bonds are made than was used to break the existing bonds are **exothermic**.

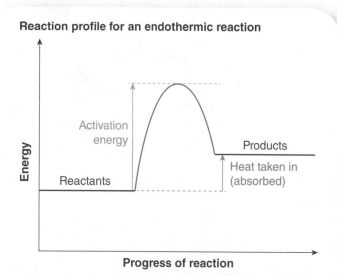

Reaction profile for an endothermic reaction

Energy / Progress of reaction — Activation energy, Reactants, Products, Heat taken in (absorbed)

Chemical reactions in which more energy is used to break the existing bonds than is released in making the new bonds are **endothermic**.

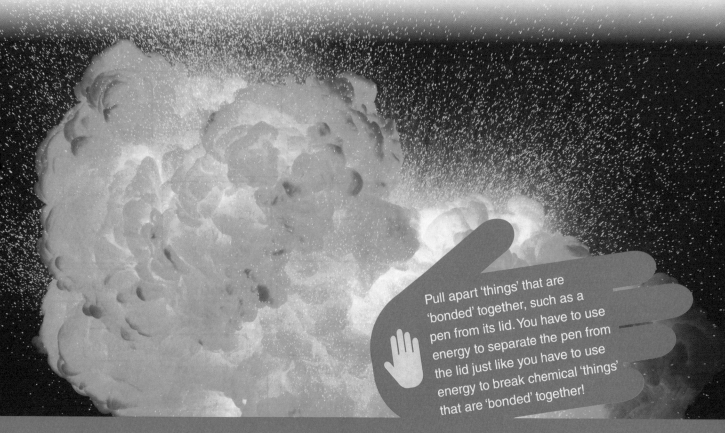

Pull apart 'things' that are 'bonded' together, such as a pen from its lid. You have to use energy to separate the pen from the lid just like you have to use energy to break chemical 'things' that are 'bonded' together!

1. What happens to the temperature of the surroundings when an endothermic reaction takes place?
2. Draw and label a reaction profile for an exothermic reaction.
3. During a reaction, more energy is used to break bonds in the reactants than is released when making the bonds in the products. Is this reaction exothermic or endothermic?

Mind map

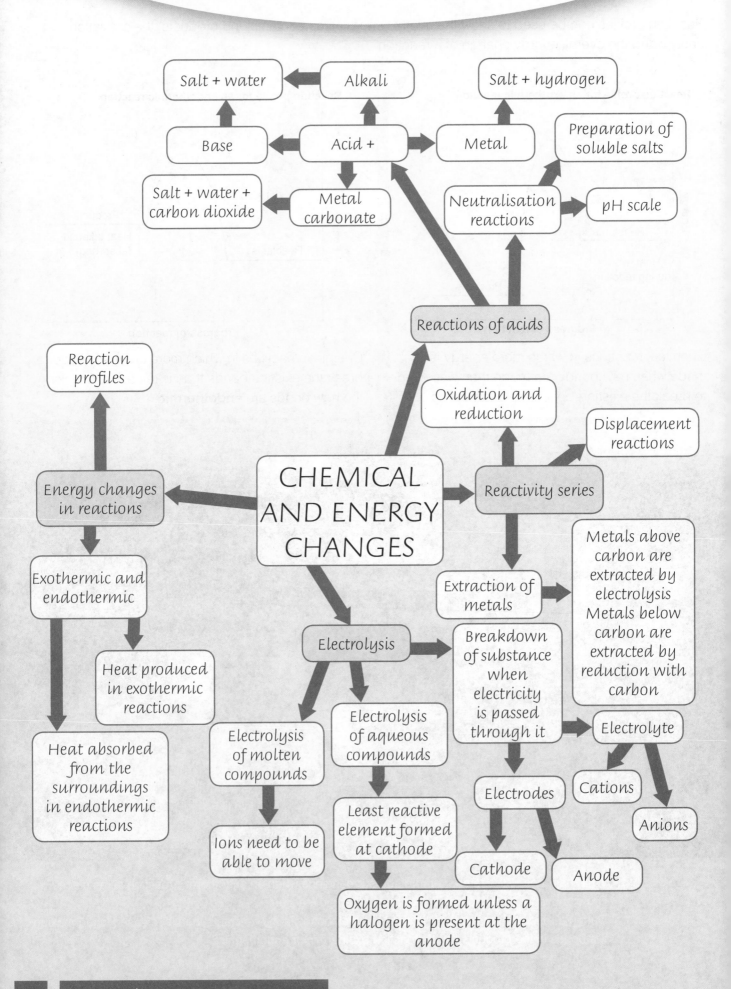

Salt + water

Alkali

Salt + hydrogen

Base

Acid +

Metal

Preparation of soluble salts

Salt + water + carbon dioxide

Metal carbonate

Neutralisation reactions

pH scale

Reactions of acids

Reaction profiles

Oxidation and reduction

Displacement reactions

Energy changes in reactions

CHEMICAL AND ENERGY CHANGES

Reactivity series

Exothermic and endothermic

Extraction of metals

Metals above carbon are extracted by electrolysis Metals below carbon are extracted by reduction with carbon

Heat produced in exothermic reactions

Electrolysis

Breakdown of substance when electricity is passed through it

Electrolyte

Heat absorbed from the surroundings in endothermic reactions

Electrolysis of molten compounds

Electrolysis of aqueous compounds

Electrodes

Cations

Ions need to be able to move

Least reactive element formed at cathode

Cathode

Anode

Anions

Oxygen is formed unless a halogen is present at the anode

Practice questions

1. Zinc is found in the Earth's crust as zinc oxide.

   a) Write a balanced symbol equation for the reaction between zinc and oxygen to form zinc oxide. **(1 mark)**

   b) Zinc can be extracted by reacting the zinc oxide with magnesium, as shown in the equation below.

   $Mg + ZnO \rightarrow MgO + Zn$

   Which species has been reduced in this reaction?

   Explain your answer. **(2 marks)**

   c) Is zinc more or less reactive than magnesium?

   Explain your answer with reference to the above equation. **(2 marks)**

   d) Would you expect zinc metal to normally be extracted from its ore by electrolysis or by reduction with carbon? Explain your answer. **(2 marks)**

2. When methane burns in air the equation for the reaction is:

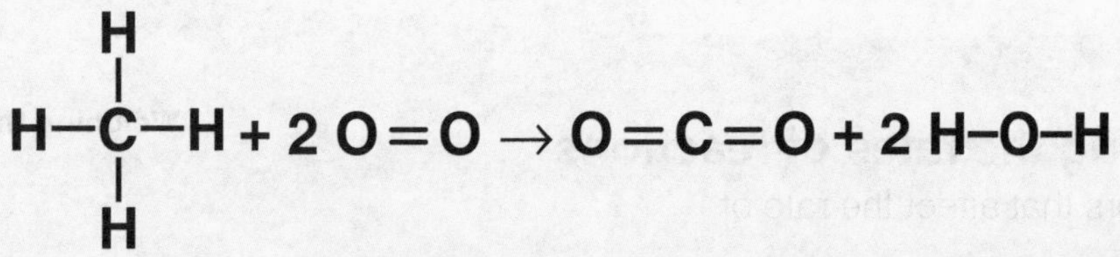

$$\underset{\begin{array}{c}|\\H\end{array}}{\overset{\begin{array}{c}H\\|\end{array}}{H-C-H}} + 2\,O{=}O \rightarrow O{=}C{=}O + 2\,H{-}O{-}H$$

   a) The reaction between methane and oxygen is exothermic.

   Draw and label a reaction profile for this reaction. **(3 marks)**

   b) During an exothermic reaction, what happens to the temperature of the surroundings? **(1 mark)**

Rates of reaction and affecting factors

54

Keywords

Rate ➤ A measure of the speed of a chemical reaction

Catalyst ➤ A species that alters the rate of a reaction without being used up or chemically changed at the end

Activation energy ➤ The minimum amount of energy that particles must collide with in order to react

Calculating rates of reactions

The **rate** of a chemical reaction can be determined by measuring the quantity of a reactant used or (more commonly) the quantity of a product formed over time.

$$\text{mean rate of reaction} = \frac{\text{quantity of reactant used}}{\text{time taken}}$$

$$\text{mean rate of reaction} = \frac{\text{quantity of product formed}}{\text{time taken}}$$

For example, if 46 cm³ of gas is produced in 23 seconds then the mean rate of reaction is 2 cm³/s.

Rates of reaction can also be determined from graphs.

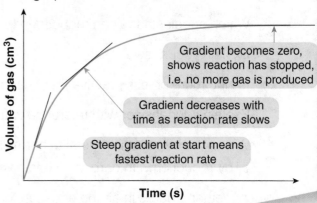

Gradient becomes zero, shows reaction has stopped, i.e. no more gas is produced

Gradient decreases with time as reaction rate slows

Steep gradient at start means fastest reaction rate

Factors affecting the rates of reactions

There are five factors that affect the rate of chemical reactions:
➤ concentrations of the reactants in solution
➤ pressure of reacting gases
➤ surface area of any solid reactants
➤ temperature
➤ presence of a **catalyst**.

During experiments the rate of a chemical reaction can be found by…
➤ measuring the mass of the reaction mixture (e.g. if a gas is lost during a reaction)
➤ measuring the volume of gas produced
➤ observing a solution becoming opaque or changing colour.

As particle size decreases ➤ The surface area to volume ratio increases ➤ and the rate of reaction increases

Weighing the reaction mixture

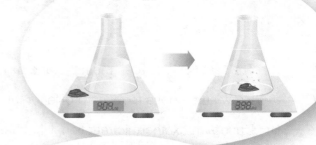

Measuring the volume of gas produced

Observing the formation of a precipitate

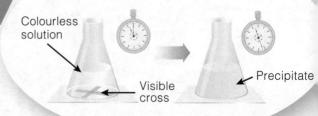

Colourless solution

Visible cross

Precipitate

Collision theory and rates of reaction

Collision theory explains how various factors affect rates of reaction.
It states that, for a chemical reaction to occur…

1 The reactant particles must collide with each other. **AND** **2** They must collide with sufficient energy – this amount of energy is known as the **activation energy**.

Surface area	Temperature	Pressure	Concentration
			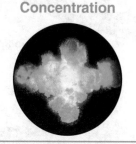
A smaller particle size means a higher surface area to volume ratio. With smaller particles, more collisions can take place, meaning a greater rate of reaction.	Increasing the temperature increases the rate of reaction because the particles are moving more quickly and so will collide more often. Also, more particles will possess the activation energy, so a greater proportion of collisions will result in a reaction.	At a higher pressure, the gas particles are closer together, so there is a greater chance of them colliding, resulting in a higher rate of reaction.	At a higher concentration, there are more reactant particles in the same volume of solution, which increases the chance of collisions and increases the rate of reaction.
Large pieces – small surface area to volume ratio  **Small pieces** – large surface area to volume ratio	**Low temperature** 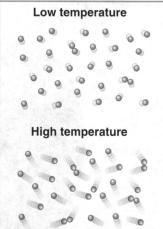 **High temperature**	**Low pressure** 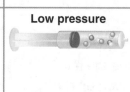 **High pressure**	**Low concentration** 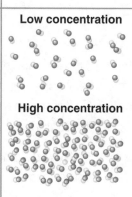 **High concentration**

1. What is the rate of the reaction in which 12 g of reactant is used up over 16 seconds?
2. State two factors that affect the rate of reaction.
3. How would you measure the rate of a reaction where one of the products is hydrogen gas?
4. What is the effect on the rate of a reaction as the surface area to volume ratio of a solid reactant increases?
5. What is collision theory?
6. What is meant by the term 'activation energy'?
7. Why does a higher concentration of solution increase the rate of a chemical reaction?

Add a teaspoon of sugar to a cup of tea or coffee and stir. Time how long it takes for the sugar to dissolve. Occasionally, lift the teaspoon to check that the sugar has all dissolved.
Make another cup of tea or coffee and repeat the experiment but using a cube of sugar instead. Does sugar dissolve more quickly as a cube or as individual granules?
Dissolving is a physical process rather than a chemical reaction but the principle is the same!

Catalysts, reversible reactions and equilibrium

55

Enzyme (biological catalyst)

Zinc catalyst

Part of car catalytic converter

Catalysts

Catalysts are chemicals that change the rate of chemical reactions but are not used up during the reaction. Different chemical reactions need different catalysts. In biological systems enzymes act as catalysts.

Catalysts work by providing an alternative reaction pathway of lower activation energy. This can be shown on a reaction profile.

Activation energy without catalyst

Activation energy with catalyst

Reactants

Products

Energy

Progress of reaction

Catalysts are not reactants and so they are not included in the chemical equation.

Keep a list of different reactions that involve catalysts or enzymes.

Reversible reactions

In some reactions, the products of the reaction can react to produce the original reactants. These reactions are called reversible reactions. For example:

➤ Heating ammonium chloride:

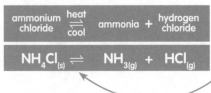

we use this symbol to represent a reversible reaction

Keyword

Equilibrium ➤ A reversible reaction where the rate of the forward reaction is the same as the rate of the reverse reaction

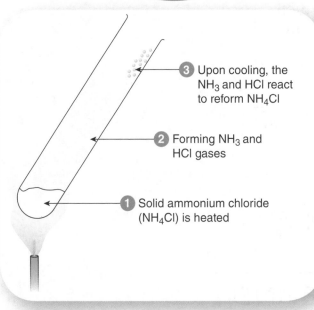

3 Upon cooling, the NH_3 and HCl react to reform NH_4Cl

2 Forming NH_3 and HCl gases

1 Solid ammonium chloride (NH_4Cl) is heated

➤ Heating hydrated copper(II) sulfate:

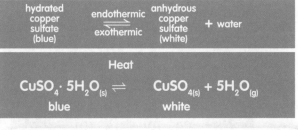

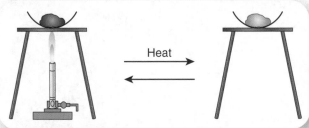

If the forward reaction is endothermic (absorbs heat) then the reverse reaction must be exothermic (releases heat).

Equilibrium

When a reversible reaction is carried out in a closed system (nothing enters or leaves) and the rate of the forward reaction is equal to the rate of the reverse reaction, the reaction is said to have reached **equilibrium**.

1. What is a catalyst?
2. Explain how catalysts increase the rate of chemical reactions.
3. Which symbol is used to represent a reversible reaction?
4. The forward reaction in a reversible reaction is exothermic. Is the reverse reaction exothermic or endothermic?

Mind map

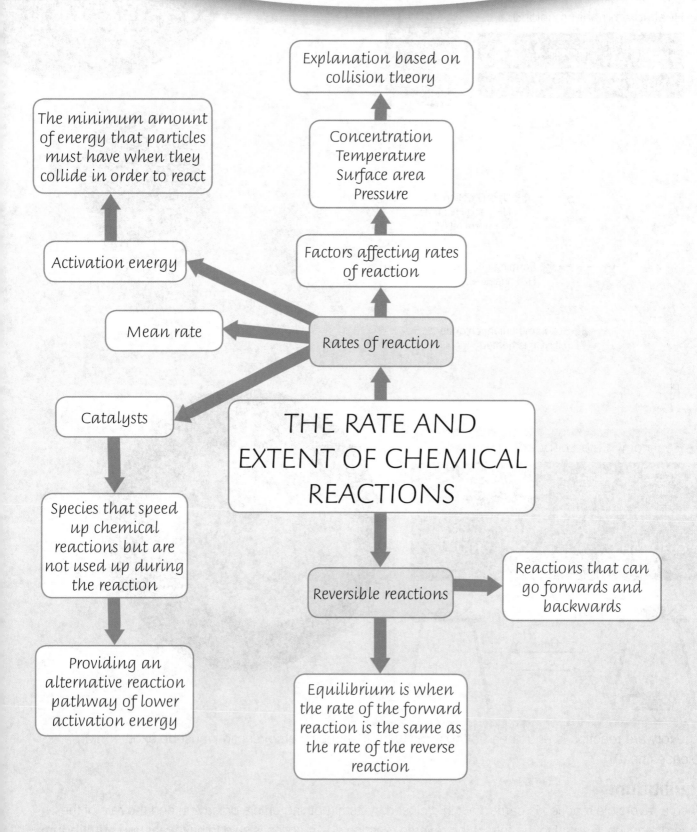

Explanation based on collision theory

The minimum amount of energy that particles must have when they collide in order to react

Concentration
Temperature
Surface area
Pressure

Activation energy

Factors affecting rates of reaction

Mean rate

Rates of reaction

Catalysts

THE RATE AND EXTENT OF CHEMICAL REACTIONS

Species that speed up chemical reactions but are not used up during the reaction

Reversible reactions

Reactions that can go forwards and backwards

Providing an alternative reaction pathway of lower activation energy

Equilibrium is when the rate of the forward reaction is the same as the rate of the reverse reaction

Practice questions

1. A student was investigating the rate of reaction between magnesium and hydrochloric acid. He measured out a 0.25 g strip of magnesium metal and then added it to 25 cm³ of acid in a conical flask. A gas syringe was attached as shown in the diagram.

   The volume of hydrogen collected was recorded every 10 seconds. A graph of the student's results is shown below.

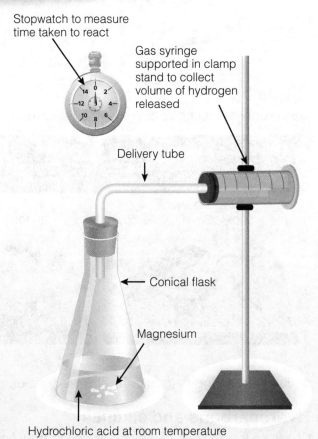

Stopwatch to measure time taken to react

Gas syringe supported in clamp stand to collect volume of hydrogen released

Delivery tube

Conical flask

Magnesium

Hydrochloric acid at room temperature

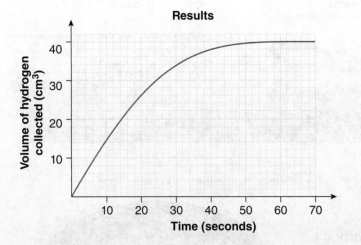

Results

   a) How long did the reaction take to finish? Explain your answer. (**2 marks**)

   b) Using your answer to part **a**, calculate the mean rate of reaction. (**2 marks**)

   c) How can you tell from the graph that the rate of reaction was faster after 10 seconds than it was after 40 seconds? (**1 mark**)

   d) The student then repeated the experiment using 0.25 g of magnesium powder instead of magnesium ribbon. Everything else was kept the same. Sketch on the graph the curve that the results from this experiment would have produced. (**2 marks**)

   e) The student predicted that carrying out the experiment with a higher concentration of acid would have increased the rate of reaction. Explain why he is correct. (**2 marks**)

2. Hydrogen for use in the Haber process or in fuel cells can be produced from methane and water according to the equation below.

$$CH_{4(g)} + H_2O_{(g)} \rightleftharpoons CO_{(g)} + 3H_{2(g)}$$

The forward reaction is endothermic.

   a) How can you tell that this process is carried out at a temperature above 100°C? (**1 mark**)

   b) What does the $\rightleftharpoons$ symbol tell you about the reaction? (**1 mark**)

Crude oil, hydrocarbons and alkanes

Keywords

Hydrocarbon ➤ A molecule containing hydrogen and carbon atoms only

Fractional distillation ➤ A method used to separate mixtures of liquids

Crude oil

This process describes how crude oil is formed.

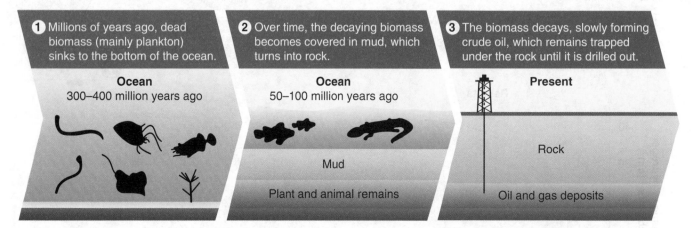

1 Millions of years ago, dead biomass (mainly plankton) sinks to the bottom of the ocean.

Ocean
300–400 million years ago

2 Over time, the decaying biomass becomes covered in mud, which turns into rock.

Ocean
50–100 million years ago

Mud

Plant and animal remains

3 The biomass decays, slowly forming crude oil, which remains trapped under the rock until it is drilled out.

Present

Rock

Oil and gas deposits

As it takes so long to form crude oil, we consider it to be a finite or non-renewable resource.

Hydrocarbons and alkanes

Crude oil is a mixture of molecules called **hydrocarbons**. Most hydrocarbons are members of a homologous series of molecules called alkanes.

Members of an homologous series...
➤ have the same general formula
➤ differ by CH_2 in their molecular formula from neighbouring compounds
➤ show a gradual trend in physical properties, e.g. boiling point
➤ have similar chemical properties.

Alkanes are hydrocarbons that have the general formula C_nH_{2n+2}.

Alkane	Methane, CH_4	Ethane, C_2H_6	Propane, C_3H_8	Butane, C_4H_{10}
Displayed formula	H | H–C–H | H	H H | | H–C–C–H | | H H	H H H | | | H–C–C–C–H | | | H H H	H H H H | | | | H–C–C–C–C–H | | | | H H H H

Make up your own mnemonic to remember the different fractions in a fractionating column.

Fractional distillation

Crude oil on its own is relatively useless. It is separated into more useful components (called fractions) by **fractional distillation**. The larger the molecule, the stronger the intermolecular forces and so the higher the boiling point.

56

Crude oil is heated until it evaporates.

It then enters a fractionating column which is hotter at the bottom than at the top

where the molecules condense at different temperatures.

Groups of molecules with similar boiling points are collected together. They are called fractions.

The fractions are sent for processing to produce fuels and feedstock (raw materials) for the petrochemical industry

which produces many useful materials, e.g. solvents, lubricants, detergents and polymers (plastics).

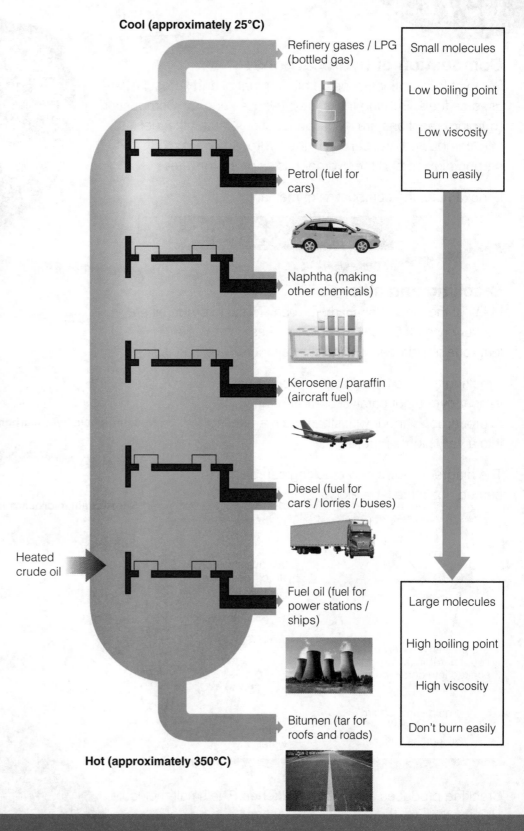

Fractionating column

Cool (approximately 25°C)

Refinery gases / LPG (bottled gas)

Petrol (fuel for cars)

Naphtha (making other chemicals)

Kerosene / paraffin (aircraft fuel)

Diesel (fuel for cars / lorries / buses)

Heated crude oil

Fuel oil (fuel for power stations / ships)

Bitumen (tar for roofs and roads)

Hot (approximately 350°C)

Small molecules

Low boiling point

Low viscosity

Burn easily

Large molecules

High boiling point

High viscosity

Don't burn easily

1. What is crude oil formed from?
2. What are hydrocarbons?
3. What is the molecular formula of propane?
4. By what process is crude oil separated?

Combustion and cracking of hydrocarbons

57

Combustion of hydrocarbons

Some of the fractions of crude oil (e.g. petrol and kerosene) are used as fuels. Burning these fuels releases energy. Combustion (burning) reactions are oxidation reactions. When the fuel is fully combusted, the carbon in the hydrocarbons is oxidised to carbon dioxide and the hydrogen is oxidised to water.

For example, the combustion of methane:

$$CH_{4(g)} + 2O_{2(g)} \rightarrow CO_{2(g)} + 2H_2O_{(l)}$$

Cracking and alkenes

Many of the long-chain hydrocarbons found in crude oil are not very useful. **Cracking** is the process of turning a long-chain hydrocarbon into shorter, more useful ones.

Cracking is done by passing hydrocarbon vapour over a hot catalyst or mixing the hydrocarbon vapour with steam before heating it to a very high temperature.

The diagram shows how cracking can be carried out in the laboratory.

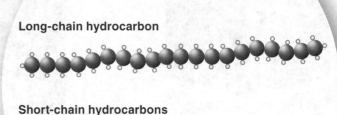

Long-chain hydrocarbon

Short-chain hydrocarbons

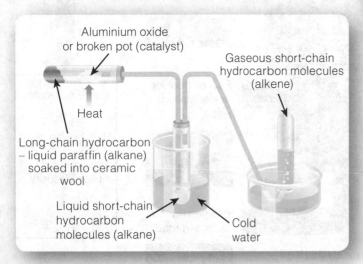

Aluminium oxide or broken pot (catalyst)

Gaseous short-chain hydrocarbon molecules (alkene)

Heat

Long-chain hydrocarbon – liquid paraffin (alkane) soaked into ceramic wool

Liquid short-chain hydrocarbon molecules (alkane)

Cold water

Cracking produces **alkanes** and **alkenes**. The small-molecule alkanes that are formed during cracking are in high demand as fuels. The alkenes (which are more reactive than alkanes) are mostly used to make plastics by the process of polymerisation and as starting materials for the production of many other chemicals.

Keywords

Cracking ➤ A process used to break up large hydrocarbon molecules into smaller, more useful molecules.
Alkane ➤ Hydrocarbons that have the general formula C_nH_{2n+2}
Alkene ➤ Hydrocarbons that contain a carbon–carbon double bond

There are many different equations that can represent cracking. This is because the long hydrocarbon can break in many different places.
A typical equation for the cracking of the hydrocarbon decane ($C_{10}H_{22}$) is:

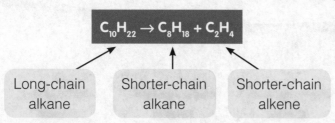

$$C_{10}H_{22} \rightarrow C_8H_{18} + C_2H_4$$

Long-chain alkane   Shorter-chain alkane   Shorter-chain alkene

Example: The cracking of dodecane ($C_{12}H_{26}$) forms one molecule of ethene (C_2H_4), one molecule of butane (C_4H_{10}) and two molecules of another hydrocarbon, as shown in the equation below:

$$C_{12}H_{26} \rightarrow C_2H_4 + C_4H_{10} + 2 \ldots\ldots\ldots\ldots$$

Complete the equation for the cracking of dodecane ($C_{12}H_{26}$) by working out the molecular formula of the other hydrocarbon formed.

There are 12 carbon atoms on the left-hand side of the equation and only 6 on the right (2 + 4). Therefore, the two molecules of hydrocarbon must contain 6 carbon atoms in total, i.e. 3 carbon atoms per molecule.

There are 26 hydrogen atoms on the left-hand side of the equation and only 14 on the right (4 + 10). Therefore, the two molecules of hydrocarbon must contain 12 hydrogen atoms in total, i.e. 6 hydrogen atoms per molecule.

Therefore, the molecular formula of the missing hydrocarbon is C_3H_6 and the completed equation is:

$$C_{12}H_{26} \rightarrow C_2H_4 + C_4H_{10} + 2C_3H_6$$

The presence of alkenes can be detected using bromine water. Alkenes decolourise bromine water but when it is mixed with alkanes, the bromine water stays orange.

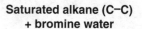 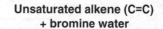

Unsaturated alkene (C=C) + bromine water   Saturated alkane (C–C) + bromine water

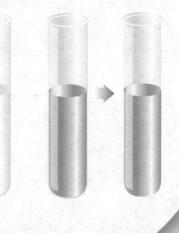

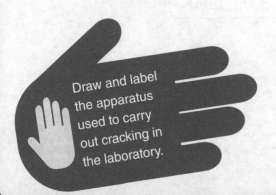

Draw and label the apparatus used to carry out cracking in the laboratory.

1. Write a balanced symbol equation for the complete combustion of ethane, C_2H_6.
2. In this equation, give the formula for the missing hydrocarbon formed by the cracking of octane: $C_8H_{18(g)} \rightarrow 2C_2H_{4\ (g)} + \underline{\hspace{3cm}}$
3. As well as using a catalyst, how else can cracking be carried out?
4. What is the chemical test and observation for alkenes?
5. What is the main use of alkenes?

Mind map

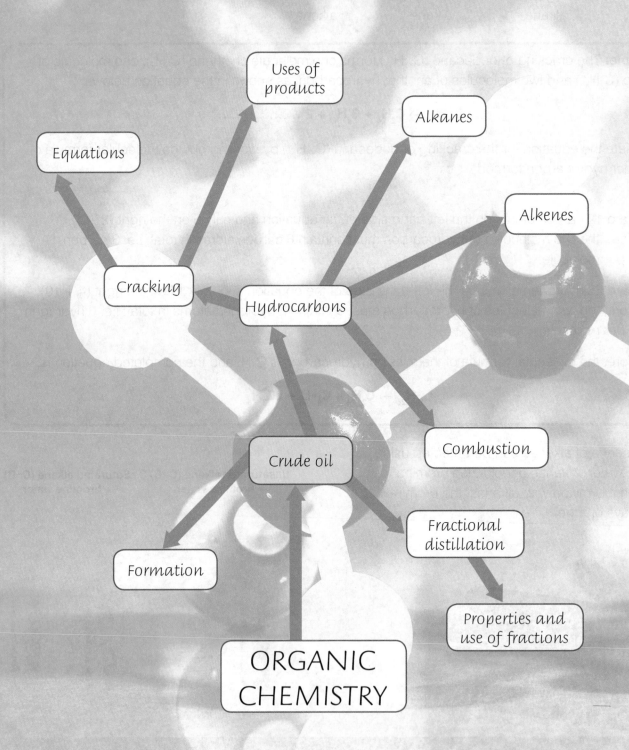

Uses of products

Equations

Alkanes

Alkenes

Cracking

Hydrocarbons

Combustion

Crude oil

Formation

Fractional distillation

Properties and use of fractions

ORGANIC CHEMISTRY

Practice questions

1. Crude oil is extracted from the earth by drilling and then it is separated into useful components. Some of the alkanes found in crude oil undergo further processing to make them more useful in a process called cracking.

   a) Describe the stages in the formation of crude oil. **(3 marks)**

   b) What is the name of the process by which crude oil is separated and by what property are the molecules in crude oil separated? **(2 marks)**

   c) What are alkanes? **(2 marks)**

   d) Why is cracking an important process? **(2 marks)**

   e) State one way in which cracking can be carried out. **(1 mark)**

   f) The hydrocarbon dodecane ($C_{12}H_{26}$) can be cracked to produce two molecules of ethene (C_2H_4) and one other molecule. Complete the equation below stating the formula of this other product.

   $$C_{12}H_{26} \rightarrow 2C_2H_4 + \text{..........................}$$ **(1 mark)**

2. Kerosene and bitumen are two fractions found in crude oil. Molecules in the kerosene fraction typically have between 11 and 18 carbon atoms. Molecules in the bitumen fraction typically have more than 35 carbon atoms.

   a) State a use for kerosene and bitumen. **(2 marks)**

   b) Which fraction, kerosene or bitumen, would you expect to be the most viscous? **(1 mark)**

   c) Which fraction, kerosene or bitumen, would you expect to be the most flammable? **(1 mark)**

   d) Which fraction, kerosene or bitumen, would you expect to have the highest boiling temperature? Explain your answer. **(2 marks)**

3. Alkanes are a homologous series of hydrocarbons with the general formula C_nH_{2n+2}.

   a) What are hydrocarbons? **(2 marks)**

   b) Other than the same general formula, give two other characteristics of members of the same homologous series. **(2 marks)**

   c) How many carbon atoms are there in the alkane containing 36 hydrogen atoms? **(1 mark)**

   d) Draw the displayed formula of the alkane containing 3 carbon atoms. **(1 mark)**

   e) Write a balanced symbol equation for the complete combustion of butane (C_4H_{10}). **(2 marks)**

   f) Another homologous series of hydrocarbons are the alkenes. Describe how you could chemically distinguish between an alkane and an alkene. **(3 marks)**

Purity, formulations and chromatography

Keywords
Formulation ➤ A mixture that has been designed as a useful product
Chromatography ➤ A method of separating mixtures of dyes

Purity

In everyday language, a pure substance can mean a substance that has had nothing added to it (i.e. in its natural state), such as milk.

In chemistry, a pure substance is a single element or compound (i.e. not mixed with any other substance).

Pure elements and compounds melt and boil at specific temperatures. For example, pure water freezes at 0°C and boils at 100°C. However, if something is added to water (e.g. salt) then the freezing point decreases (i.e. goes below 0°C) and the boiling point rises above 100°C.

Formulations

A **formulation** is a mixture that has been designed as a useful product. Many formulations are complex mixtures in which each ingredient has a specific purpose.

Formulations are made by mixing the individual components in carefully measured quantities to ensure that the product has the correct properties.

Fertilisers

Fuels

Cleaning materials

Examples of formulations

Foods

Medicines

Paints

Find five items at home or in a supermarket that are described as 'pure'. Look at their composition (e.g. from a food label) and decide whether they are chemically pure.

Chromatography

Chromatography is used to separate mixtures of dyes. It is used to help identify substances.

In paper chromatography, a solvent (the mobile phase) moves up the paper (the stationary phase) carrying different components of the mixture different distances, depending on their attraction for the paper and the solvent. In thin layer chromatography (TLC), the stationary phase is a thin layer of an inert substance (e.g. silica) supported on a flat, unreactive surface (e.g. a glass plate).

In the chromatogram on the right, substance X is being analysed and compared with samples A, B, C, D and E.

It can be seen from the chromatogram that substance X has the same pattern of spots as sample D. This means that sample X and sample D are the same substance. Pure compounds (e.g. compound A) will only produce one spot on a chromatogram.

The ratio of the distance moved by the compound to the distance moved by the solvent is known as its R_f value.

$$R_f = \frac{\text{distance moved by substance}}{\text{distance moved by solvent}}$$

Different compounds have different R_f values in different solvents. This can be used to help identify unknown compounds by comparing R_f values with known substances.

In this case, the R_f value is 0.73 $\left(\frac{4.0}{5.5}\right)$

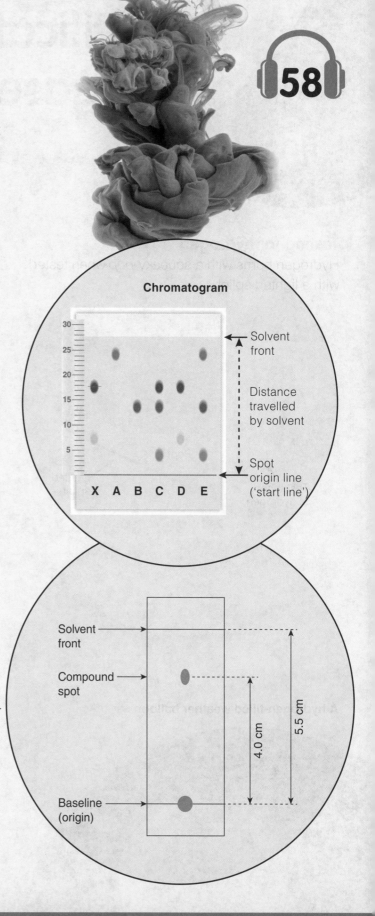

1. In chemistry what is meant by the term 'pure substance'?
2. What is a formulation?
3. Give two examples of substances that are formulations.
4. In a chromatogram, the solvent travelled 6 cm and a dye travelled 3.3 cm. What is the R_f value of the dye?

Identification of gases

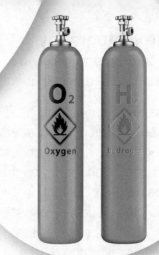

Testing for hydrogen

Hydrogen burns with a squeaky *pop* when tested with a lighted splint.

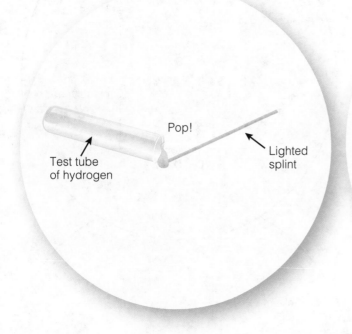

Pop!

Test tube of hydrogen

Lighted splint

Testing for oxygen

Oxygen relights a glowing splint.

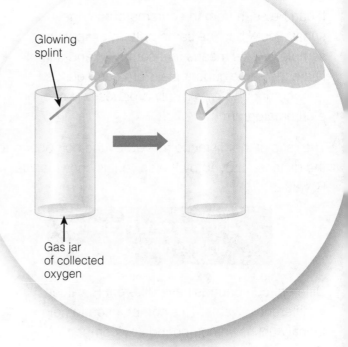

Glowing splint

Gas jar of collected oxygen

A hydrogen-filled weather balloon

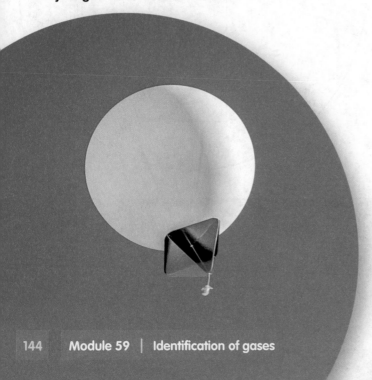

Oxygen is used in breathing masks

Carbon dioxide is released when fossil fuels are burnt

Testing for carbon dioxide

When carbon dioxide is mixed with or bubbled through limewater (calcium hydroxide solution) the limewater turns milky (cloudy).

Testing for chlorine

Chlorine turns moist blue litmus paper red before bleaching it and turning it white.

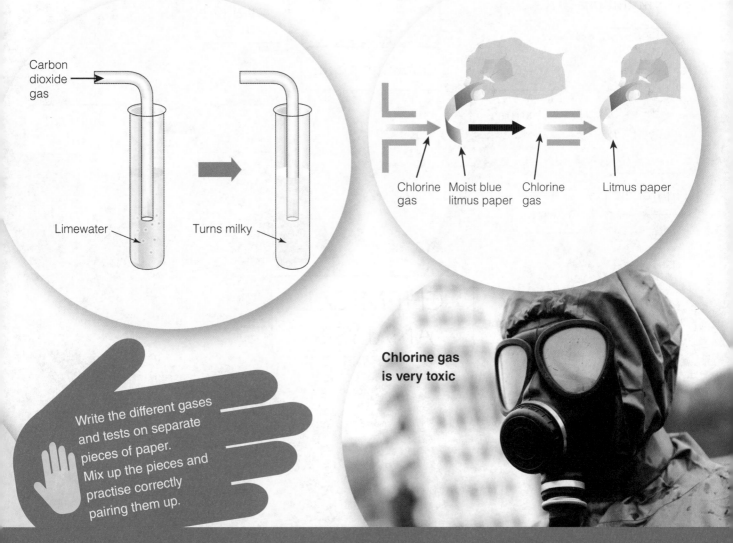

Carbon dioxide gas

Limewater

Turns milky

Chlorine gas

Moist blue litmus paper

Chlorine gas

Litmus paper

Chlorine gas is very toxic

Write the different gases and tests on separate pieces of paper. Mix up the pieces and practise correctly pairing them up.

1. Which gas relights a glowing splint?
2. What is the test for hydrogen gas?
3. What is the chemical name for limewater?
4. What happens to limewater when it is mixed with carbon dioxide gas?
5. What effect does chlorine gas have on moist blue litmus paper?

Mind map

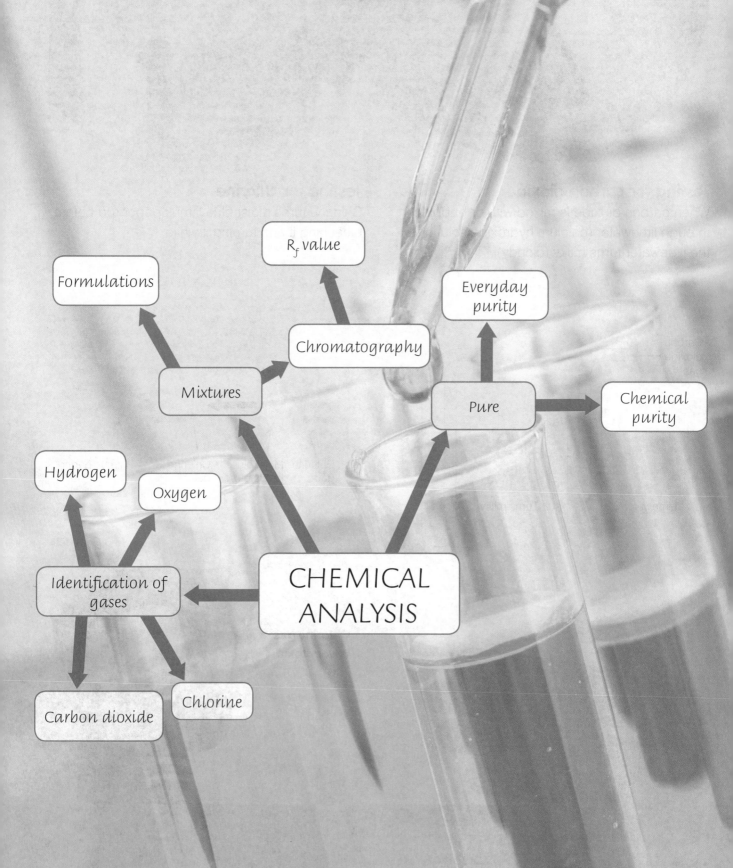

Formulations

R$_f$ value

Everyday purity

Chromatography

Mixtures

Pure

Chemical purity

Hydrogen

Oxygen

Identification of gases

CHEMICAL ANALYSIS

Carbon dioxide

Chlorine

Practice questions

1. Shown below is a chromatogram for four dyes (W, X, Y and Z) and three coloured inks (blue, red and yellow). Each ink is made up of a single compound.

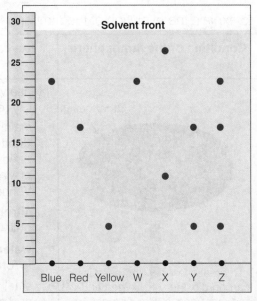

a) Which one of the dyes contains three colours?
   Explain your answer. **(2 marks)**

b) Which dyes contain blue ink? **(1 mark)**

c) Which two coloured inks are present in dye Y? **(1 mark)**

d) Which dye doesn't contain blue, red or yellow ink?
   Explain your answer. **(2 marks)**

e) Calculate the R_f value for the blue ink. **(2 marks)**

f) Which dye is chemically pure?
   Explain your answer. **(2 marks)**

g) Many industrial dyes are formulations. What is meant by the term 'formulation'? **(1 mark)**

2. Complete the table below. **(8 marks)**

Gas	Test	Observation
		A 'squeaky pop' is heard
Oxygen		
	Add moist blue litmus paper	
	Add limewater	

The Earth's early atmosphere

Theories about the composition of Earth's early atmosphere and how the atmosphere was formed have changed over time. Evidence for the early atmosphere is limited because the Earth is approximately 4.6 billion years old.

Evolution of the atmosphere

The table below gives one theory to explain the evolution of the atmosphere.

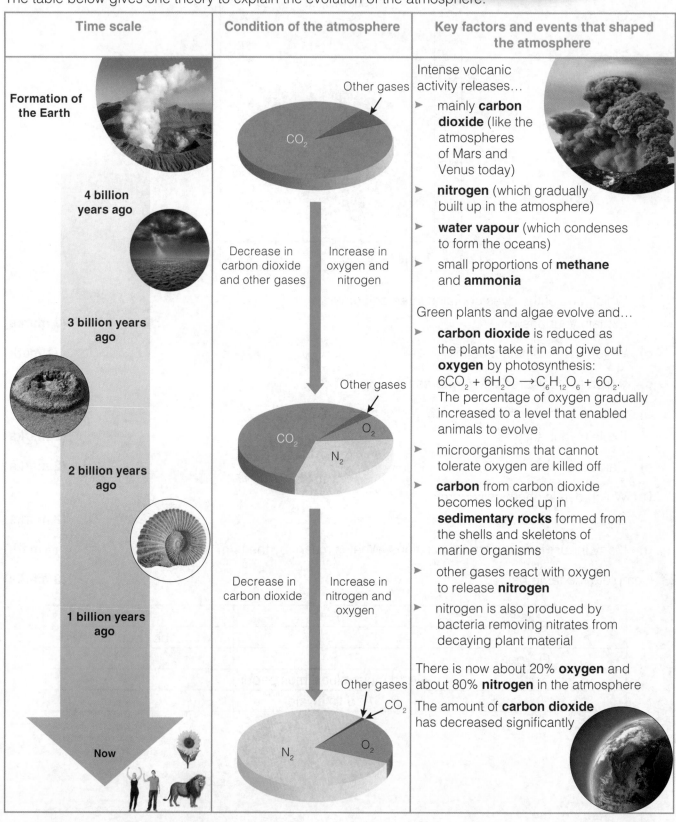

Time scale	Condition of the atmosphere	Key factors and events that shaped the atmosphere
Formation of the Earth	Other gases / CO_2	Intense volcanic activity releases… ▸ mainly **carbon dioxide** (like the atmospheres of Mars and Venus today) ▸ **nitrogen** (which gradually built up in the atmosphere) ▸ **water vapour** (which condenses to form the oceans) ▸ small proportions of **methane** and **ammonia**
4 billion years ago **3 billion years ago**	Decrease in carbon dioxide and other gases / Increase in oxygen and nitrogen Other gases / CO_2 / O_2 / N_2	Green plants and algae evolve and… ▸ **carbon dioxide** is reduced as the plants take it in and give out **oxygen** by photosynthesis: $6CO_2 + 6H_2O \rightarrow C_6H_{12}O_6 + 6O_2$. The percentage of oxygen gradually increased to a level that enabled animals to evolve ▸ microorganisms that cannot tolerate oxygen are killed off ▸ **carbon** from carbon dioxide becomes locked up in **sedimentary rocks** formed from the shells and skeletons of marine organisms ▸ other gases react with oxygen to release **nitrogen** ▸ nitrogen is also produced by bacteria removing nitrates from decaying plant material
2 billion years ago **1 billion years ago** **Now**	Decrease in carbon dioxide / Increase in nitrogen and oxygen Other gases / CO_2 / N_2 / O_2	There is now about 20% **oxygen** and about 80% **nitrogen** in the atmosphere The amount of **carbon dioxide** has decreased significantly

Composition of the atmosphere today

The proportions of gases in the atmosphere have been more or less the same for about 200 million years. **Water vapour** may also be present in varying quantities (0–3%).

Mainly argon, plus other noble gases (1%)

Carbon dioxide, CO_2 (0.03%)

Oxygen, O_2 (21%)

Nitrogen, N_2 (78%)

Keyword

Photosynthesis ➤ The process by which green plants and algae use water and carbon dioxide to make glucose and oxygen

How carbon dioxide decreased

The amount of carbon dioxide in the atmosphere today is much less than it was when the atmosphere first formed. This is because…

Make a timeline showing the key changes in the evolution of the Earth's atmosphere.

➤ green plants and algae use carbon dioxide for **photosynthesis**

➤ fossil fuels such as oil (see Module 56) and coal (a sedimentary rock made from thick plant deposits that were buried and compressed over millions of years) have captured CO_2.

➤ carbon dioxide is used to form sedimentary rocks, e.g. limestone

1. Name two gases that volcanoes released into the early atmosphere.
2. How did oxygen become present in the atmosphere?
3. How much of the atmosphere today is nitrogen gas?
4. State two ways that the amount of carbon dioxide present in the early atmosphere decreased.

Climate change

(ws) ## Human activity and global warming

Some human activities increase the amounts of greenhouse gases (water vapour, carbon dioxide and methane) in the atmosphere including...

Combustion of fossil fuels releases carbon dioxide into the atmosphere

Increased animal farming releases more methane into the atmosphere, e.g. as a by-product of digestion and decomposition of waste

Deforestation reduces the amount of carbon dioxide removed from the atmosphere by photosynthesis

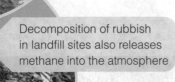

Decomposition of rubbish in landfill sites also releases methane into the atmosphere

Greenhouse gases

The temperature on Earth is maintained at a level to support life by the greenhouse gases in the atmosphere. These gases allow short wavelength radiation from the Sun to pass through but absorb the long wavelength radiation reflected back from the ground trapping heat and causing an increase in temperature. Common greenhouse gases are water vapour, carbon dioxide and methane.

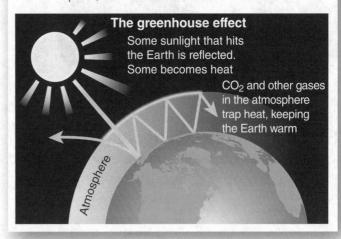

The greenhouse effect

Some sunlight that hits the Earth is reflected. Some becomes heat

CO_2 and other gases in the atmosphere trap heat, keeping the Earth warm

Atmosphere

The increase in carbon dioxide levels in the last century or so correlates with the increased use of fossil fuels by humans.

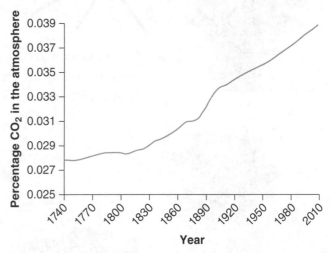

Based on **peer-reviewed evidence**, many scientists believe that increasing these human activities will lead to global climate change.

Predicting the impact of changes on global climate change is not easy because of the many different contributing factors involved. This can lead to simplified models and speculation often presented in the media that may not be based on all of the evidence. This means that the information could be biased.

Increased human activity resulting in more release of fossil fuels ⟩ Increased temperatures ⟩ Global climate change

Global climate change

Increasing average global temperature is a major cause of climate change. Potential effects of climate change include:

Rising sea levels, which may cause flooding and coastal erosion

More frequent and severe storms

Changes to the amount, timing and distribution of rainfall

Temperature and water stress for humans and wildlife

Changes in the food producing capacity of some regions

Changes to the distribution of wildlife species

NS Reducing the carbon footprint

The **carbon footprint** is a measure of the total amount of carbon dioxide (and other greenhouse gases) emitted over the life cycle of a product, service or event.

Problems of trying to reduce the carbon footprint include…

➤ disagreement over the causes and consequences of global climate change
➤ lack of public information and education
➤ lifestyle changes, e.g. greater use of cars / aeroplanes
➤ economic considerations, i.e. the financial costs of reducing the carbon footprint
➤ incomplete international co-operation.

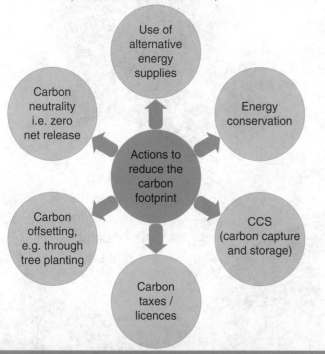

Use of alternative energy supplies

Carbon neutrality i.e. zero net release

Energy conservation

Actions to reduce the carbon footprint

Carbon offsetting, e.g. through tree planting

CCS (carbon capture and storage)

Carbon taxes / licences

Produce a leaflet informing members of the public about how they can reduce their own carbon footprint.

Keywords

Peer-reviewed evidence ➤ Work (evidence) of scientists that has been checked by other scientists to ensure that it is accurate and scientifically valid

Carbon footprint ➤ The total amount of carbon dioxide (and other greenhouse gases) emitted over the full life cycle of a product, service or event

1. Name two greenhouse gases.
2. State one way that human activity leads to an increased amount of methane in the atmosphere.
3. Give two potential effects of global climate change.
4. Give an example of one problem of trying to reduce the carbon footprint.

61

Atmospheric pollution

62

Pollutants from fuels

The combustion of **fossil fuels** is a major source of atmospheric pollutants. Most fuels contain carbon and often sulfur is present as an impurity. Many different gases are released into the atmosphere when a fuel is burned.

Solid particles and unburned hydrocarbons can also be released forming **particulates** in the air.

Gases produced by burning fuels

- carbon monoxide
- carbon dioxide
- oxides of nitrogen
- water vapour
- sulfur dioxide

Sulfur dioxide is produced by the oxidation of sulfur present in fuels – often from coal-burning power stations

Carbon monoxide and soot (carbon) are produced by incomplete combustion of fuels

Oxides of nitrogen are formed from the reaction between nitrogen and oxygen from the air – often from the high temperatures and sparks in the engines of motor vehicles

Keywords

Fossil fuel ➤ Fuel formed in the ground over millions of years from the remains of dead plants and animal

Particulates ➤ Small solid particles present in the air

Properties and effects of atmospheric pollutants

Carbon monoxide is a colourless, odourless toxic gas and so it is difficult to detect. It combines with haemoglobin in the blood, which reduces the oxygen-carrying capacity of blood.

Sulfur dioxide and **oxides of nitrogen** cause respiratory problems in humans and can form acid rain in the atmosphere. Acid rain damages plants and buildings.

Particulates in the atmosphere can cause global dimming, which reduces the amount of sunlight that reaches the Earth's surface. Breathing in particulates can also damage lungs, which can cause health problems.

Make a list of all of the ways in one day that you see waste gases being released into the atmosphere.

1. Name two gases produced by burning fuels.
2. How is carbon monoxide formed?
3. What problems do atmospheric sulfur dioxide and oxides of nitrogen cause?

Using the Earth's resources and obtaining potable water

Keywords

Sustainable development ➤ Living in a way that meets the needs of the current generation without compromising the potential of future generations to meet their own needs

Potable water ➤ Water that is safe to drink

Earth's resources

We use the Earth's **resources** to provide us with warmth, shelter, food and transport. These needs are met from natural resources which, supplemented by agriculture, provide food, timber, clothing and fuels. Resources from the earth, atmosphere and oceans are processed to provide energy and materials.

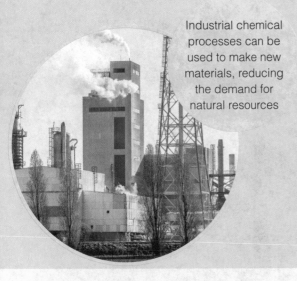

Industrial chemical processes can be used to make new materials, reducing the demand for natural resources

Chemistry plays a role in providing **sustainable development**. This means that the needs of the current generation are met without compromising the potential of future generations to meet their own needs. For example...

Chemistry plays an important role in improving agricultural processes, e.g. by developing fertilisers

Drinking water

Water that is safe to drink is called **potable water**. It is not pure in the chemical sense (see Module 58) because it contains dissolved minerals and ions.

Water of appropriate quality is essential for life. This means that it contains sufficiently low levels of dissolved salts and microbes.

In the UK, most potable water comes from rainwater. To turn rainwater into potable water, water companies carry out a number of processes.

① An appropriate source of fresh water is selected

② It is then passed through filter beds to remove any solid impurities

③ Finally it is sterilised (suitable sterilising agents include chlorine, ozone and ultraviolet light) to kill microbes, making it safe to drink

Waste water

Urban lifestyles and industrial processes generate large quantities of waste water, which requires treatment before being released into the environment. Sewage and agricultural waste water require removal of organic matter and harmful microbes. Industrial waste water may require removal of organic matter and harmful chemicals.

Sewage treatment includes...

➤ screening and grit removal
➤ sedimentation to produce sewage sludge and effluent
➤ anaerobic digestion of sewage sludge
➤ aerobic biological treatment of effluent.

When supplies of fresh water are limited, removal of salt (desalination) of salty water / seawater can be used.

This is done in two ways:

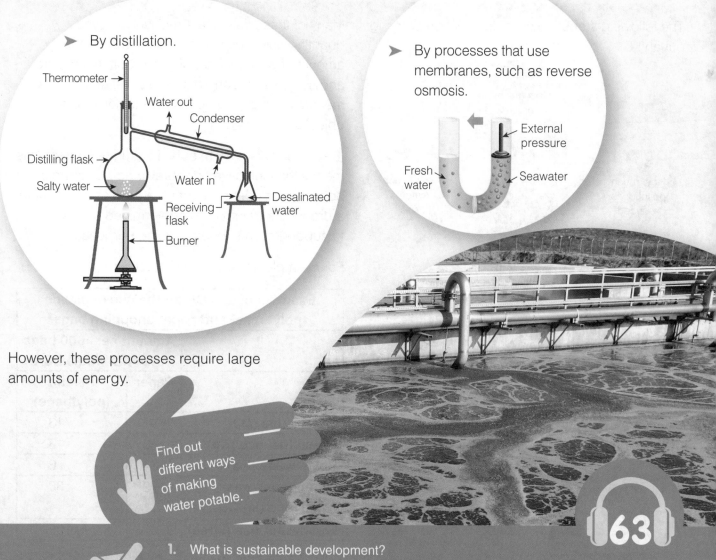

➤ By distillation.

Thermometer
Water out
Condenser
Distilling flask
Water in
Salty water
Desalinated water
Receiving flask
Burner

➤ By processes that use membranes, such as reverse osmosis.

External pressure
Fresh water
Seawater

However, these processes require large amounts of energy.

Find out different ways of making water potable.

63

1. What is sustainable development?
2. State the difference between pure water and potable water.
3. What are the two main stages in turning rainwater into potable water?
4. Name one way of desalinating salty water.

Life-cycle assessment and recycling

WS Life-cycle assessments

Life-cycle assessments (LCAs) are carried out to evaluate the environmental impact of products in each of the following stages.

| Extracting and processing raw materials | Manufacturing and packaging | Disposal at the end of useful life | Transport and distribution at each of the previous stages |

The following steps are considered when carrying out an LCA.

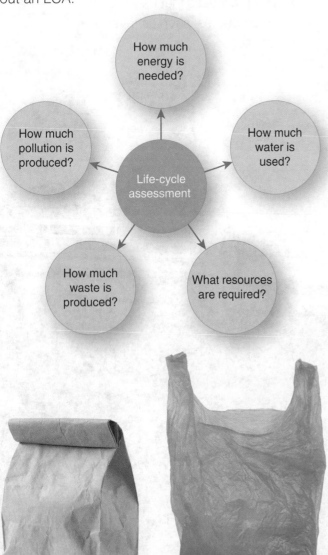

- How much energy is needed?
- How much water is used?
- What resources are required?
- How much waste is produced?
- How much pollution is produced?

Life-cycle assessment

Many of these values are relatively easy to quantify. However, some values, such as the amount of pollution, are often difficult to measure and so value judgements have to be made. This means that carrying out an LCA is not a purely objective process.

It is not always easy to obtain accurate figures. This means that selective or abbreviated LCAs, which are open to bias or misuse, can be devised to evaluate a product, to reinforce predetermined conclusions or to support claims for advertising purposes.

For example, look at the LCA below.

Example of an LCA for the use of plastic (polythene) and paper shopping bags		
	Amount per 1000 bags over the whole LCA	
	Paper	**Plastic (polythene)**
Energy use (MJ)	2590	713
Fossil fuel use (Kg)	28	13
Solid waste (Kg)	34	6
Greenhouse gas emissions (kg CO_2)	72	36
Freshwater use (litres)	3387	198

This LCA provides evidence supporting the argument that using polythene bags is better for the environment than paper bags!

Ways of reducing the use of resources

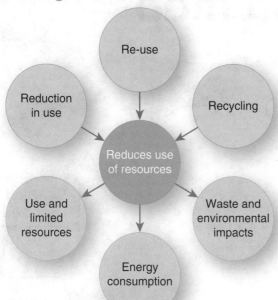

- Re-use
- Reduction in use
- Recycling
- Reduces use of resources
- Use and limited resources
- Waste and environmental impacts
- Energy consumption

Keywords

Life-cycle assessment ➤ An evaluation of the environmental impact of a product over the whole of its lifespan
Blast furnace ➤ Industrial method of extracting iron from iron ore

Quarrying

Many materials such as glass, metals, building materials, plastics and clay ceramics are produced from limited raw materials. Most of the energy used in their production comes from limited resources, such as fossil fuels. Obtaining raw materials from the earth by quarrying and mining has a detrimental environmental impact.

Some products, like glass, can be **reused** (e.g. washing and then using again for the same purpose). Recycled glass is crushed, melted and remade into glass products.

Recycling

Other products cannot be reused and so they are **recycled** for a different use.

Metals are recycled by sorting them, followed by melting them and recasting / reforming them into different products. The amount of separation required for recycling depends on the material (e.g. whether they are magnetic or not) and the properties required of the final product. For example, some scrap steel can be added to iron from a **blast furnace** to reduce the amount of iron that needs to be extracted from iron ore.

Blast furnace

Try to identify the environmental impact of this book by considering the effects of all of the stages of its life cycle.

1. What does a life-cycle assessment measure?
2. State two factors that a life-cycle assessment tries to evaluate.
3. Why are life-cycle assessments not always totally objective?
4. State one way in which we can reduce the use of resources.

Mind map

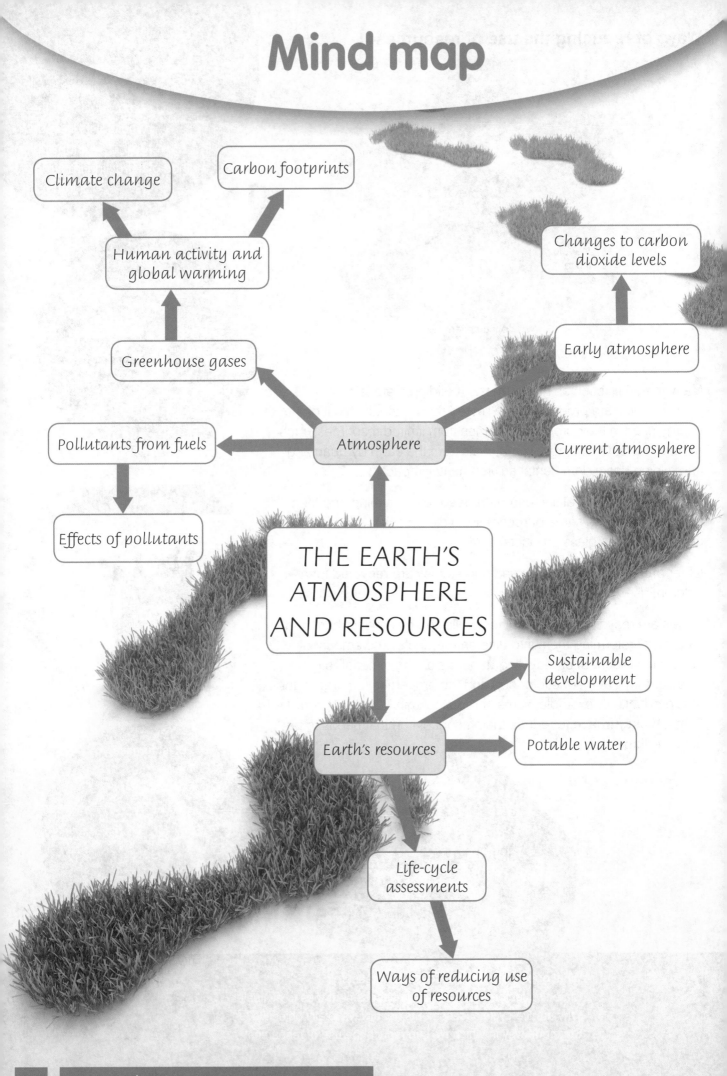

Climate change

Carbon footprints

Human activity and global warming

Changes to carbon dioxide levels

Greenhouse gases

Early atmosphere

Pollutants from fuels

Atmosphere

Current atmosphere

Effects of pollutants

THE EARTH'S ATMOSPHERE AND RESOURCES

Sustainable development

Earth's resources

Potable water

Life-cycle assessments

Ways of reducing use of resources

Practice questions

1. The composition of the Earth's atmosphere today is very different from how it was 4.6 million years ago. The pie charts below show the composition of the atmosphere today compared with millions of years ago.

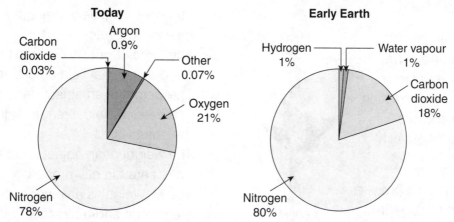

   a) When the Earth was formed, where did many of the gases in the early atmosphere come from? **(1 mark)**

   b) Name one gas whose composition has remained fairly constant over time. **(1 mark)**

   c) Explain why oxygen is now present in the atmosphere and why it has not always been there. **(2 marks)**

   d) Why is the amount of carbon dioxide in the atmosphere much lower today than in the early atmosphere? **(2 marks)**

   e) Explain why the level of carbon dioxide in the atmosphere has been steadily increasing over the last couple of centuries. **(2 marks)**

2. This question is about climate change and the carbon footprint.

   a) Other than combustion of fossil fuels, describe **two** ways that human activity increases the amount of greenhouse gases in the atmosphere. **(2 marks)**

   b) What is meant by 'carbon footprint'? **(2 marks)**

   c) Outline **two** actions that can be taken to reduce the carbon footprint. **(2 marks)**

   d) As well as carbon dioxide and water vapour, the combustion of fossil fuels produces many other atmospheric pollutants.

      i) Other than carbon dioxide, name one other gas produced by the combustion of fossil fuels. **(2 marks)**

      ii) Outline the effect on the environment that the gas you have named in part i) has on the environment. **(1 mark)**

65

Forces

Forces

Scalar quantities have magnitude only.

Vector quantities have magnitude and an associated direction.

A vector quantity can be represented by an arrow. The length of the arrow represents the magnitude, and the direction of the arrow represents the direction of the vector quantity.

600N

600N

At **terminal velocity** the downwards force on the skydiver (weight) is equal to the upwards force (air resistance). As these forces have the same magnitude the vector arrows representing them are the same length.

Contact and non-contact forces

A force is a push or pull that acts on an object due to the interaction with another object. Force is a vector quantity. All forces between objects are either:

➤ contact forces – the objects are physically touching, for example:
friction, air resistance, tension and normal contact force

or

➤ non-contact forces – the objects are physically separated, for example:
gravitational force, electrostatic force and magnetic force.

Friction

Air resistance

Gravity

➤ **Weight** is the force acting on an object due to gravity.
➤ All matter has a gravitational field that causes attraction. The field strength is much greater for massive objects.
➤ The force of gravity close to the Earth is due to the gravitational field around the Earth.
➤ The weight of an object depends on the gravitational field strength at the point where the object is.
➤ The weight of an object and the mass of an object are directly proportional (weight $\propto$ mass). Weight is a vector quantity as it has a magnitude and a direction. Mass is a scalar quantity as it only has a magnitude.
➤ The weight of an object can be thought of acting on a single point called the 'centre of mass'.

Weight is measured using a calibrated spring-balance – a **newtonmeter**.

Weight can be calculated using the following equation:

> **weight = mass × gravitational field strength**
> $W = mg$
> ➤ weight, W, in newtons, N
> ➤ mass, m, in kilograms, kg
> ➤ gravitational field strength, g, in newtons per kilogram, N/kg

Example:
What is the weight of an object with a mass of 54 kg in a gravitational field strength of 10 N/kg?

weight = mass × gravitational field strength
     = 54 kg × 10 N/kg
     = 540 N

Resultant force

The resultant force equals the total effect of all the different forces acting on an object.

Work done and energy transfer

Work is done when a force causes an object to move. The force causes a displacement.

The work done by a force on an object can be calculated using the following equation:

---

work done = force × distance moved along the line of action of the force

$$W = Fs$$

➤ work done, *W*, in joules, J

➤ force, *F*, in newtons, N

➤ distance, *s*, in metres, *m* (*s* represents displacement, commonly called distance)

Example:

What work is done when a force of 90 N moves an object 14 m?

work done = force × distance moved along the line of action of the force

work done = 90 × 14 = 1260 J

One joule of work is done when a force of one newton causes a displacement of one metre.

$$1 \text{ joule} = 1 \text{ newton metre}$$

Work done against the frictional forces acting on an object causes a rise in the temperature of the object.

Keywords

Scalar ➤ A quantity that only has a magnitude

Vector ➤ A quantity that has both a magnitude and a direction

Weight ➤ Force acting on an object due to gravity

 Create a comic strip that contains examples of each of the contact and non-contact forces described on the opposite page.

1. What is the difference between a vector quantity and a scalar quantity?
2. Give two examples of contact forces.
3. What is the weight of a 67 g object in a gravitational field strength of 10 N/kg?
4. What is 78 Nm in joules?

Forces and elasticity

Elastically deformed

Inelastically deformed

When an object is stretched and returns to its original length after the force is removed, it is **elastically deformed**.

When an object does not return to its original length after the force has been removed, it is **inelastically deformed**. This is **plastic deformation**.

Extension

The extension of an elastic object, such as a spring, is directly proportional to the force applied (extension ∝ force applied), if the limit of proportionality is not exceeded.

➤ **Stretching** – when a spring is stretched, the force pulling it exceeds the force of the spring.

➤ **Bending** – when a shelf bends under the weight of too many books, the downward force is from the weight of the books, and the shelf is resisting this weight.

➤ **Compressing** – when a car goes over a bump in the road, the force upwards is opposed by the springs in the car's suspension.

In order for any of these processes to occur, there must be more than one force applied to the object to bring about the change.

The force on a spring can be calculated using the following equation:

force = spring constant × extension
$$F = ke$$
➤ force, F, in newtons, N
➤ spring constant, k, in newtons per metre, N/m
➤ extension, e, in metres, m

Example:
A spring with spring constant 35 N/m is extended by 0.3 m. What is the force on the spring?

$F = ke$
$\quad = 35 \times 0.3$
$\quad = 10.5$ N

This relationship also applies to the compression of an elastic object, where the extension e would be the compression of the object.

A force that stretches (or compresses) a spring does **work** and **elastic potential energy** is stored in the spring. Provided the spring does not go past the limit of proportionality, the work done on the spring equals the stored elastic potential energy.

Spring balance

Force and extension

Force and extension have a linear relationship.

If force and extension are plotted on a graph, the points can be connected with a straight line (see Graph 1).

The points in a non-linear relationship cannot be connected by a straight line (see Graph 2).

Keywords

Elastically deformed ➤ Stretched object that returns to its original length after the force is removed
Inelastically deformed ➤ Stretched object that does not return to its original length after the force is removed

Graph 1 – Linear relationship

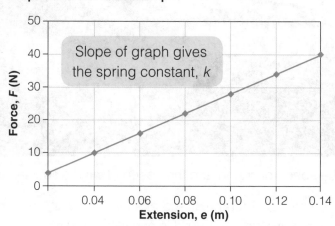

Slope of graph gives the spring constant, *k*

Graph 2 – Non-linear relationship

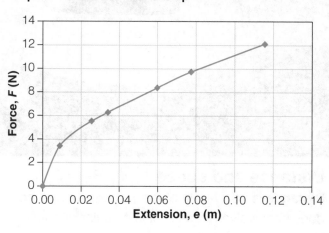

In order to calculate the spring constant in Graph 1, take two points and apply the equation.

$$\text{So, spring constant} = \frac{\text{force}}{\text{extension}}$$

$$= \frac{10}{0.04}$$

$$= 250 \text{ N/m}$$

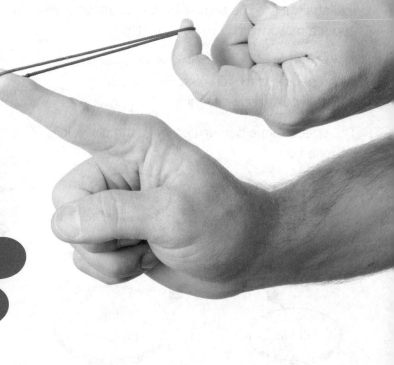

Attach a number of objects of different masses to an elastic band and measure how far the band extends. Repeat the investigation again using an plastic bag instead of a rubber band. How do your results compare?

1. What is the force of a spring with an extension of 0.1 m and a spring constant of 2 N/m?
2. When is the extension of an elastic object, such as a spring, directly proportional to the force applied?
3. What type of energy is stored in a spring?

66

Speed and velocity

67

Distance and speed

Distance is how far an object moves. As distance does not involve direction, it is a scalar quantity.

Displacement includes both the distance an object moves, measured in a straight line from the start point to the finish point, and the direction of that straight line. As displacement has magnitude and a direction, it is a vector quantity.

The speed of a moving object is rarely constant. When people walk, run or travel in a car, their speed is constantly changing.

The speed that a person can walk, run or cycle depends on:

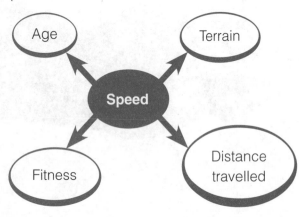

Some typical speeds are:
- ➤ walking – 1.5 m/s
- ➤ running – 3 m/s
- ➤ cycling – 6 m/s
- ➤ car – 20 m/s
- ➤ train – 35 m/s

The speed of sound and the speed of wind also vary.

A typical value for the speed of sound in air is 330 m/s. A gale force wind is one with a speed above 14 m/s.

For an object travelling at a constant speed, distance travelled can be calculated by the following equation:

distance travelled = speed × time

$$s = vt$$

➤ distance, s, in metres, m

➤ speed, v, in metres per second, m/s

➤ time, t, in seconds, s

Example:

What distance is covered by a runner with a speed of 3 m/s in 6000 seconds?

distance travelled = speed × time

= 3 × 6000

= 18 000 m

= 1.8 km

The equation can be rearranged to find speed and time.

$$\text{speed} = \frac{\text{distance travelled}}{\text{time}}$$

$$\text{time} = \frac{\text{distance travelled}}{\text{speed}}$$

Example 1:

What is the speed of a car that travels 120m in 4 seconds?

$$\text{speed} = \frac{\text{distance travelled}}{\text{time}}$$

$$= \frac{120}{4}$$

$$= 30 \text{ m/s}$$

Example 2:

How long does it take a train to travel 560m if it's travelling at 35m/s?

$$\text{time} = \frac{\text{distance travelled}}{\text{speed}}$$

$$= \frac{560}{35}$$

$$= 16 \text{ m/s}$$

Keywords

Displacement ➤ Distance travelled in a given direction

Velocity ➤ Speed in a given direction

Velocity

The **velocity** of an object is its speed in a given direction. As velocity has a magnitude and a direction, it is a vector quantity.

Time how long it takes to travel to school. Use a mapping website to calculate the distance. Now calculate the average speed of your journey. Replicate this investigation on different days. Is there much variation in your average speed?

1. Why is distance a scalar quantity?
2. What is the speed of a walker who covers 10 km in 2.5 hours?
3. What is the difference between speed and velocity?

Distance–time and velocity–time graphs

68

Distance and time

The distance an object moves in a straight line can be represented by a distance–time graph.

The speed of an object can be calculated from the gradient of its distance–time graph.

The graph below shows a person jogging.

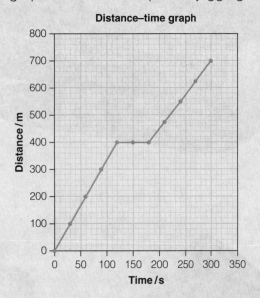

Distance–time graph

The average speed of this jogger between 0–120 seconds can be worked out as follows:

$$\text{speed} = \frac{\text{distance}}{\text{time}} = \frac{400}{120} = 3.33 \text{ m/s}$$

The speed of an accelerating object can be determined by using a tangent to measure the gradient of the distance–time graph.

Acceleration

Acceleration can be calculated using the following equation:

$$\text{average acceleration} = \frac{\text{change in velocity}}{\text{time taken}}$$

$$\left[a = \frac{\Delta v}{t} \right]$$

➤ acceleration, a, in metres per second squared, m/s^2
➤ change in velocity, Δv, in metres per second, m/s
➤ time, t, in seconds, s

Example:
A car travels from 0 to 27 m/s in 3.8 seconds. What is its acceleration?

$$\text{change in velocity} = 27 - 0 = 27 \text{ m/s}$$
$$\text{average acceleration} = \frac{\text{change in velocity}}{\text{time taken}}$$
$$= \frac{27}{3.8} = 7.1 \text{ m/s}^2$$

An object that slows down (decelerates) has a negative acceleration.

The acceleration of a runner who goes from 0 m/s to 3 m/s in 5 seconds is $\frac{(3 - 0)}{5} = 0.6 \text{ m/s}^2$

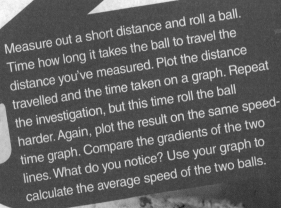

Measure out a short distance and roll a ball. Time how long it takes the ball to travel the distance you've measured. Plot the distance travelled and the time taken on a graph. Repeat the investigation, but this time roll the ball harder. Again, plot the result on the same speed-time graph. Compare the gradients of the two lines. What do you notice? Use your graph to calculate the average speed of the two balls.

Velocity–time graph

Acceleration can be calculated from the gradient of a velocity–time graph.

Keyword

Acceleration ➤ Change in velocity over time

The graph below shows the movement of a car.

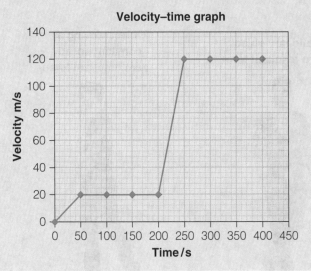

Velocity–time graph

The acceleration between 0–50 seconds can be worked out as follows:

$$\text{gradient of line} = \frac{\text{velocity}}{\text{time}}$$

$$= \frac{20}{50} = 0.4 \text{ m/s}^2$$

As the gradient of the line is steeper between 200–250 seconds than it is between 0–50 seconds, the acceleration must have been greater between 200–250 seconds.

The equation below applies to uniform motion:

final velocity² − initial velocity² = 2 × acceleration × distance

$$v^2 - u^2 = 2\,as$$

➤ final velocity, v, in metres per second, m/s
➤ initial velocity, u, in metres per second, m/s
➤ acceleration, a, in metres per second squared, m/s²
➤ distance, s, in metres, m

Falling objects

Near the Earth's surface, any object falling freely under gravity has an acceleration of about 9.8 m/s². This is the gravitational field strength, or acceleration due to gravity, and is used to calculate weight.

An object falling through a fluid initially accelerates due to the force of gravity. Eventually the resultant force will be zero and the object will move at its terminal velocity.

1. What does the gradient of a distance–time graph represent?
2. What type of acceleration will an object which is slowing down have?

69

Newton's laws

Newton's first law

Newton's first law deals with the effect of resultant forces.

➤ If the resultant force acting on an object is zero and the object is stationary, the object remains stationary.

➤ If the resultant force acting on an object is zero and the object is moving, the object continues to move at the same speed and in the same direction (its velocity will stay the same).

➤ When a vehicle travels at a steady (constant) speed the resistive forces balance the driving force.

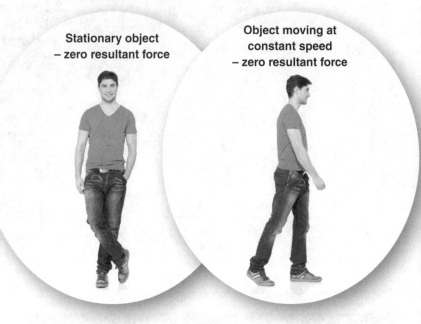

Stationary object – zero resultant force

Object moving at constant speed – zero resultant force

Newton's first law means that the velocity of an object will only change if a resultant force is acting on the object.

Examples of resultant force:

A car is stationary. At this point the resultant force is zero.

The car starts to move and accelerates. As it is accelerating, the resultant force on the car is no longer zero.

The car travels at a constant velocity. The resultant force is zero again.

Newton's second law

The acceleration of an object is proportional to the resultant force acting on the object, and inversely proportional to the mass of the object.

Therefore:

> **acceleration ∝ resultant force**
>
> **resultant force = mass × acceleration**
> - ➤ force, *F*, in newtons, N
> - ➤ mass, *m*, in kilograms, kg
> - ➤ acceleration, *a*, in metres per second squared, m/s^2

Example:
A motorbike and rider of mass 270 kg accelerate at 6.7 m/s^2. What is the resultant force on the motorbike?

$$270 \times 6.7 = 1809 \text{ N}$$

If the motorbike maintains a constant speed (e.g. 27 m/s) the resultant force would now be 0 (as acceleration = 0, $270 \times 0 = 0$).

Newton's third law

Whenever two objects interact, the forces they exert on each other are equal and opposite.

When a fish swims it exerts a force on the water, pushing it backwards. The water exerts an equal and opposite force on the fish, pushing it forwards.

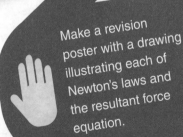

Make a revision poster with a drawing illustrating each of Newton's laws and the resultant force equation.

1. A book of weight 67 N is on a desk. What force is the desk exerting on the book?
2. A sprinter of mass 89 kg accelerates at 10 m/s^2. What is the resultant force?
3. What happens to the speed of a moving object that has a resultant force of 0 acting on it?

70

Forces and braking

Stopping distance

The stopping distance of a vehicle is a sum of the **thinking distance** and the **braking distance**.

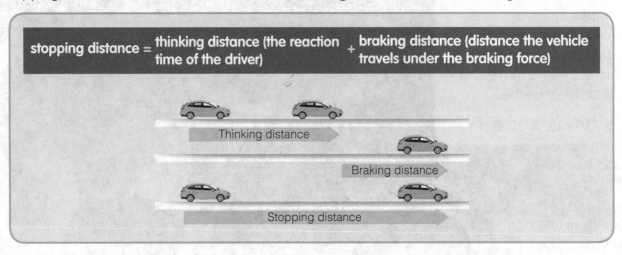

stopping distance = thinking distance (the reaction time of the driver) + braking distance (distance the vehicle travels under the braking force)

Thinking distance

Braking distance

Stopping distance

A greater vehicle speed leads to a greater stopping distance (given a set braking force). As the vehicle is going faster, the car will travel further in the time taken for the driver to react and apply the brakes.

As the mass of a vehicle increases the braking distance also increases. Thinking distance remains the same. This means that a bus would have a longer braking distance than a car and therefore a longer stopping distance.

The graph below shows the stopping distances over a range of speeds for a car.

Typical stopping distances

Speed	Thinking	Braking	Total
20 mph (32 km/h)	6 m	6 m	12 metres / 3 car lengths
30 mph (48 km/h)	9 m	14 m	23 metres / 6 car lengths
40 mph (64 km/h)	12 m	24 m	36 metres / 9 car lengths
50 mph (80 km/h)	15 m	38 m	53 metres / 13 car lengths
60 mph (96 km/h)	18 m	55 m	73 metres / 18 car lengths
70 mph (112 km/h)	21 m	75 m	96 metres / 24 car lengths

Key   thinking distance   braking distance

Design a road safety poster summarising stopping distance, thinking distance, braking distance and all the factors that affect them.

Braking distance

The braking distance of a vehicle can be affected by adverse road and weather conditions and poor condition of the vehicle.

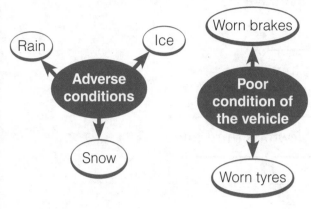

Keywords

Keywords

Thinking distance ➤ Distance travelled before the driver applies the brakes

Braking distance ➤ Distance a vehicle travels after the brakes have been applied

Braking force

This is what happens when a force is applied to the brakes of a vehicle.

| Work is done by the friction force between the brakes and the wheel. | This reduces the kinetic energy of the vehicle. | The temperature of the brakes increases. |

Reaction time

The average reaction time is between 0.2–0.9 seconds but this varies from person to person. Thinking distance increases as speed increases. This is because a faster vehicle travels further during the time taken to react.

A driver's reaction time can be affected by tiredness, drugs and alcohol. Distractions, such as eating, drinking or using a mobile phone, may also affect a driver's ability to react.

A range of different methods can be used to measure reaction times. One example is using a computer programme that flashes a colour or sounds a noise. The participant must then press a key on the keyboard as fast as possible. Average reaction times for this test are often around 0.2–0.3 seconds.

➤ The greater the speed of a vehicle, the greater the braking force needed to stop the vehicle in a certain distance.

➤ The greater the braking force, the greater the deceleration of the vehicle.

➤ Large decelerations may lead to brakes overheating and / or loss of control.

1. What two distances make up the stopping distance?
2. Give two examples of adverse road conditions which can increase the stopping distance.

Mind map

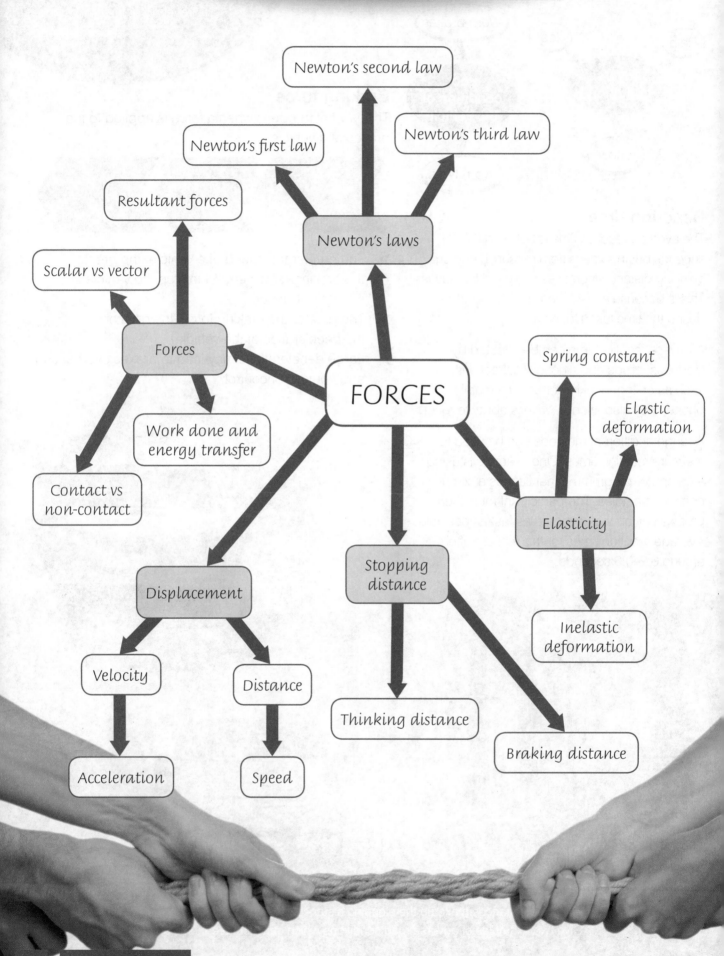

Newton's second law

Newton's first law

Newton's third law

Newton's laws

Resultant forces

Scalar vs vector

Forces

FORCES

Spring constant

Elastic deformation

Work done and energy transfer

Elasticity

Contact vs non-contact

Displacement

Stopping distance

Inelastic deformation

Velocity

Distance

Thinking distance

Braking distance

Acceleration

Speed

Practice questions

1. **a)** The gravitational field strength on Mars is 3.7 N/kg.

    What would be the weight of a Mars rover with a mass of 187 kg? **(2 marks)**

    b) Would the weight of the rover be different on Earth? Explain your answer. **(2 marks)**

2. The graph shows the velocity of an object.

    a) Use the graph to calculate the:

    i) acceleration from 0–3 seconds. **(2 marks)**

    ii) distance travelled in 3 seconds. **(2 marks)**

    b) Was the acceleration between 0–3 seconds greater than the acceleration between 3–6 seconds? Explain your answer. **(2 marks)**

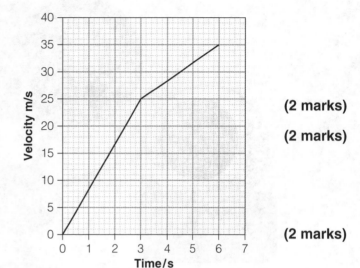

3. **a)** The diagram on the right shows the forces acting on a stationary bike and rider.

    What is the upward force from the ground? Explain how you arrived at your answer. **(2 marks)**

    b) The diagram below shows the same bike and rider moving.

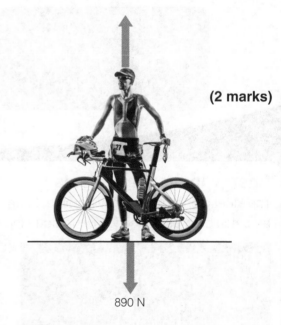

25 N ← → 25 N

890 N

   Is the rider accelerating, decelerating or moving at the same speed? Explain, using Newton's laws, how you arrived at your answer. **(2 marks)**

    c) A change in terrain caused the friction to increase. What effect would this have on the resultant force and the speed of the rider? **(2 marks)**

4. A car in a crash test has a total mass of 1200 kg. The car is travelling at a speed of 16 m/s.

    The car accelerates at 2.5 m/s². What is its resultant force? **(2 marks)**

71

Changes in energy

A system is an object or group of objects. When a system changes, the way energy is stored in it also changes.

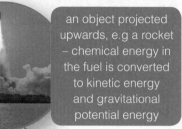

an object projected upwards, e.g a rocket – chemical energy in the fuel is converted to kinetic energy and gravitational potential energy

bringing water to a boil in an electric kettle – electrical energy to heat and sound energy

a moving object hitting an obstacle – kinetic energy to sound, heat and kinetic energy

Changes

an object accelerated by a constant force, e.g a tennis player hitting a tennis ball – chemical energy in the player to kinetic energy in their arm to kinetic energy in the racket and then kinetic energy in the ball

a vehicle slowing down – kinetic energy to sound and heat energy (friction)

Energy in moving objects

The kinetic energy of a moving object can be calculated using the following equation:

kinetic energy = 0.5 × mass × (speed)²

$$E_k = \frac{1}{2}mv^2$$

➤ kinetic energy, E_k, in joules, J

➤ mass, m, in kilograms, kg

➤ speed, v, in metres per second, m/s

Example:
A ball of mass 0.3 kg falls at 10 m/s. What is the ball's kinetic energy?

$$E_k = \frac{1}{2}mv^2$$
$$= 0.5 \times 0.3 \times 10^2$$
$$= 15 \text{ J}$$

Create a large formula poster with each of the formulas on these pages, a worked example different to the ones on this page and a diagram showing the energy transfer.

Elastic potential energy (E_e)

elastic potential energy $= 0.5 \times$ spring constant $\times$ extension²

$$E_e = \frac{1}{2}ke^2$$

(assuming the limit of proportionality has not been exceeded)

> elastic potential energy, E_e, in joules, J
> spring constant, k, in newtons per metre, N/m
> extension, e, in metres, m

Example:
A spring has a spring constant of 500 N/m and is extended by 0.2 m. What is the elastic potential energy in the spring?

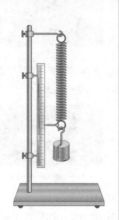

$$E_e = \frac{1}{2}ke^2$$
$$= \frac{1}{2} \, 500 \times 0.2^2$$
$$= 250 \times 0.04$$
$$= 10 \text{ J}$$

Gravitational potential energy (g.p.e.)

The amount of gravitational potential energy gained by an object raised above ground level can be calculated using the following equation:

g.p.e. = mass × gravitational field strength × height

$$E_p = mgh$$

> gravitational potential energy, E_p, in joules, J
> mass, m, in kilograms, kg
> gravitational field strength, g, in newtons per kilogram, N/kg
> height, h, in metres, m

Example:
A car of mass 1500 kg drives up a hill which is 67 m high. What is the gravitational potential energy of the car at the top of the hill? (Assume the gravitational field strength is 10 N/kg)

$$E_p = mgh$$
$$= 15\,000 \times 67 \times 10$$
$$= 10\,050 \text{ kJ}$$

Changes in thermal energy

The **specific heat capacity** of a substance is the amount of energy required to raise the temperature of one kilogram of the substance by one degree Celsius. The specific heat capacity can be used to calculate the amount of energy stored in or released from a system as its temperature changes. This can be calculated using the following equation:

change in thermal energy = mass × specific heat capacity × temperature change

$$\Delta E = m\,c\,\Delta\theta$$

> change in thermal energy, ΔE, in joules, J
> mass, m, in kilograms, kg
> specific heat capacity, c, in joules per kilogram per degree Celsius, J/kg °C
> temperature change, $\Delta\theta$, in degrees Celsius, °C

Example:
0.75 kg of 100°C water cools to 23°C. What is the change in thermal energy?
(Specific heat capacity of water = 4184 J/kg °C)

$$\Delta E = mc\,\Delta\theta$$
$$= 0.75 \times 4184 \times (100 - 23)$$
$$= 241\,626 \text{ J} = 242 \text{ kJ}$$

1. What is the kinetic energy of a 65 g object travelling at 6 m/s?
2. What is the gravitational potential energy of a 17 kg object at a height of 987 m? (gravitational field strength = 10 g)
3. What is specific heat capacity?

Conservation and dissipation of energy

Keyword

Thermal insulation ➤ Material used to reduce transfer of heat energy

Energy transfers in a system

Energy can be **transferred**, **stored** or **dissipated**. It cannot be created or destroyed.

In a closed system there is no net change to the total energy when energy is transferred.

Only part of the energy is usefully transferred. The rest of the energy dissipates and is transferred in less useful ways, often as heat or sound energy. This energy is considered **wasted**.

Examples of processes which cause a rise in temperature and so waste energy as heat include:
➤ friction between the moving parts of a machine

If the energy that is wasted can be reduced, that means more energy can be usefully transferred. The less energy wasted, the more efficient the transfer.

Efficiency

The energy efficiency for any energy transfer can be calculated using the following equation:

$$\text{efficiency} = \frac{\text{useful output energy transfer}}{\text{useful input energy transfer}}$$

Efficiency may also be calculated using the following equation:

$$\text{efficiency} = \frac{\text{useful power}}{\text{total power output}}$$

Example:

Kat uses a hairdryer. Some of the energy is wasted as sound. The electrical energy input is 24 kJ. The energy wasted is 7 kJ. What is the efficiency of the hairdryer?

First calculate the useful energy transferred.

useful energy = total energy − wasted energy
= 24 − 7 = 17 kJ

Now calculate the efficiency.

$$\text{efficiency} = \frac{\text{useful output energy transfer}}{\text{useful input energy transfer}}$$

$$= \frac{17}{24} = 0.71$$

This decimal efficiency can be represented as a percentage by multiplying it by 100.

$$0.71 \times 100 = 71\%$$

There are many ways energy efficiency can be increased:
➤ Lubrication, thermal insulation and low resistance wires reduce energy waste and improve efficiency.
➤ Thermal insulation, such as loft insulation, reduces heat loss.

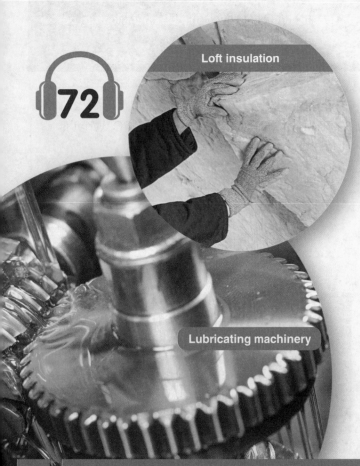

72

Loft insulation

Lubricating machinery

Power

Power is defined as the rate at which energy is transferred or the rate at which work is done. The power equation is:

$$\text{power} = \frac{\text{work done}}{\text{time}} \qquad P = \frac{W}{T}$$

➤ P = power (watts)
➤ E = work done (joules)
➤ T = time (seconds)

An energy transfer of 1 joule per second is equal to a power of 1 watt.

Example:

A weightlifter is lifting weights that have a mass of 80 kg. What power is required to lift them 2 metres vertically in 4 seconds? (Assume gravitational field strength of 10 N/kg)

$$\text{power} = \frac{\text{work done}}{\text{time}}$$

work done = force × distance
force = mass × gravitational field strength
= 80 × 10 = 800 N

work done = 800 × 2 = 1600 J

$$\text{power} = \frac{1600}{4} = 800 \text{ W}$$

A second weightlifter lifted the same mass to the same height in 3 seconds. As he carried out the same amount of work but in a shorter time, he would have a greater power.

Record all the energy that is wasted from appliances you use in a day. What do you think are the most wasteful appliances? (Hint: they're likely to be the ones that are noisy or get hot.)

Thermal insulation

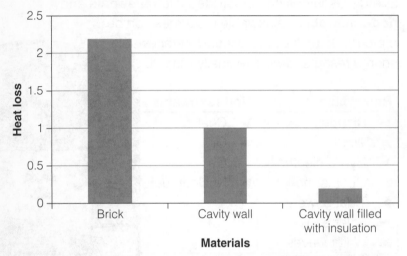

Thermal insulation has a low thermal conductivity, so has a slow rate of energy transfer by conduction. U-values give a measure of the heat loss through a substance. A higher U-value indicates that a material has a higher thermal conductivity.

Changing the material of walls to materials that have a lower thermal conductivity reduces heat loss (the U-value) and so a building cools more slowly, reducing heating costs. The graph below shows the heat lost by different types of wall.

Bar chart — Y axis: Heat loss (0 to 2.5); X axis: Materials (Brick ≈ 2.2, Cavity wall ≈ 1.0, Cavity wall filled with insulation ≈ 0.2)

Increasing the thickness of walls could also reduce heat loss.

The house below is not fitted with insulation, so lots of heat is being lost through the walls due to their high thermal conductivity. This can be seen by the red/orange colour on the infrared image.

1. Why is it important to increase the efficiency of an energy transfer?
2. A washing machine transfers 400 J of useful energy out of a total of 652 J. What is the efficiency of the washing machine?
3. What effect does fitting cavity wall insulation have on the thermal conductivity of the building's wall?

National and global energy resources

Keywords

Renewable ➤ Energy resource that can be replenished as it is used

Non-renewable ➤ Energy resource that will eventually run out

Energy resources

Energy resources are used for transport, electricity generation and heating.

Energy resources can be divided into **renewable** and **non-renewable**. Renewable resources can be replenished as they are used. Non-renewable energy resources will eventually run out.

Renewable	Non-renewable
➤ Bio-fuel	➤ Coal
➤ Wind	➤ Oil
➤ Hydro-electricity	➤ Gas
➤ Geothermal	➤ Nuclear fuel
➤ Tidal power	
➤ Solar power	
➤ Water waves	

Non-renewable

Renewable

73

Reliability

Fossil fuels (coal, oil and gas) and nuclear fuel are very reliable as they can always be used to release energy. Fossil fuels are burnt to release the stored chemical energy, and nuclear fuel radiates energy.

The wind and the Sun are examples of energy resources that are not very reliable, as the wind doesn't always blow and it's not always sunny.

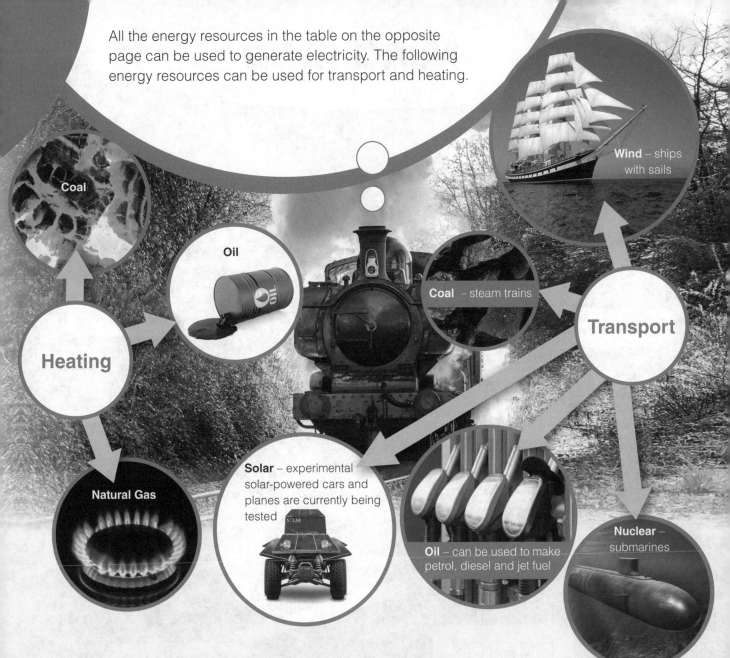

All the energy resources in the table on the opposite page can be used to generate electricity. The following energy resources can be used for transport and heating.

Coal

Oil

Wind – ships with sails

Coal – steam trains

Transport

Heating

Natural Gas

Solar – experimental solar-powered cars and planes are currently being tested

Oil – can be used to make petrol, diesel and jet fuel

Nuclear – submarines

Ethical and environmental concerns

➤ Burning fossil fuels produces carbon dioxide, which is a greenhouse gas. Increased greenhouse gas emissions are leading to climate change. Particulates and other pollutants are also released, which cause respiratory problems.

➤ Nuclear power produces hazardous nuclear waste and also has the potential for nuclear accidents, with devastating health and environmental consequences.

➤ Some people consider wind turbines to be ugly and to spoil the landscape.

➤ Building tidal power stations can lead to the destruction of important tidal habitats.

In recent years, there have been attempts to move away from over-reliance on fossil fuels to renewable forms of energy. However, fossil fuels are still used to deliver the vast majority of our energy needs.

Write out each of the different energy resources on cards, then mix the cards up and sort them into two piles: renewable and non-renewable.

1. Give three examples of renewable energy resources and three examples of non-renewable energy resources.
2. Why aren't solar and wind power reliable energy resources?
3. What is an environmental concern about burning fossil fuels?

Mind map

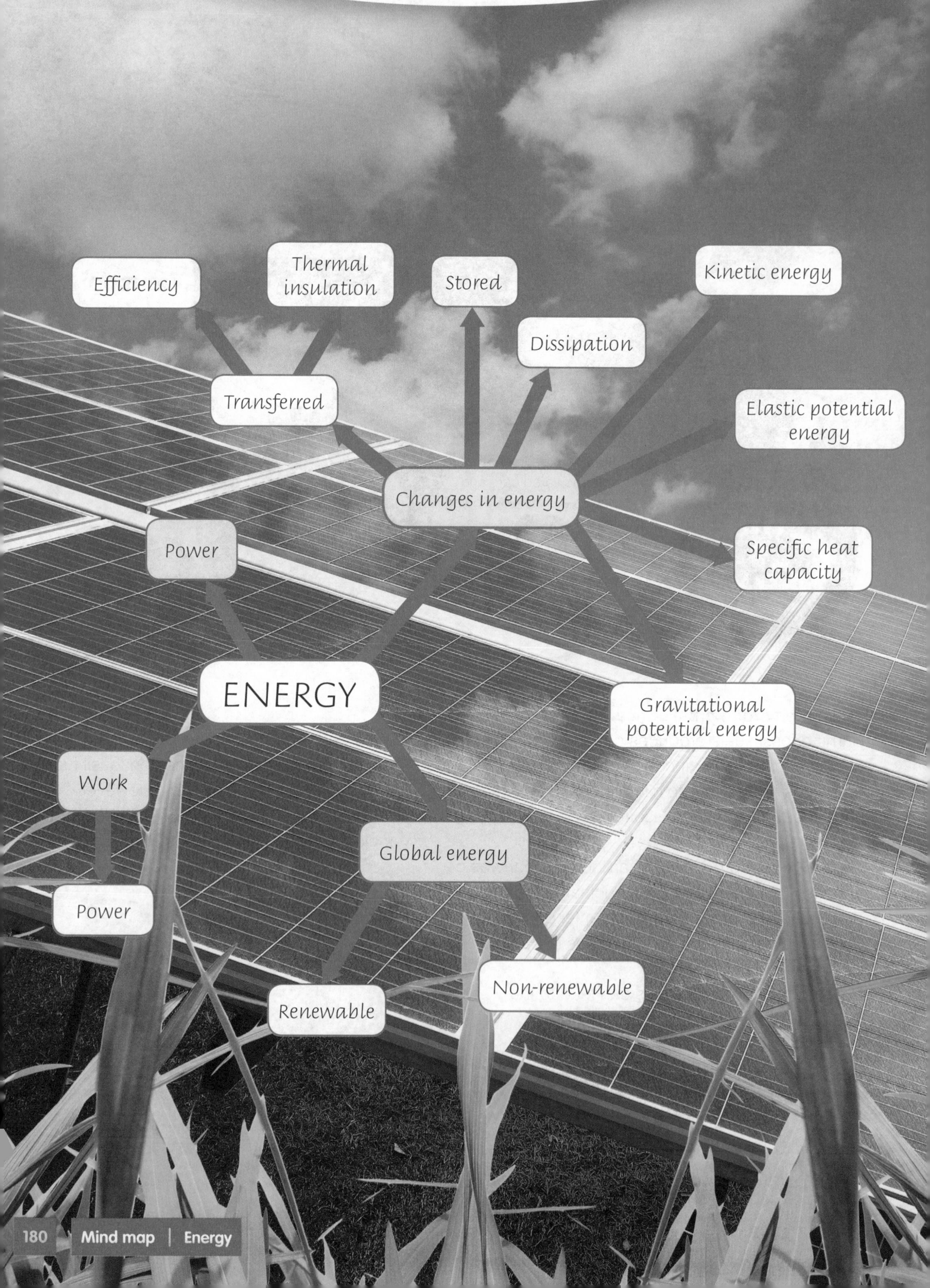

Efficiency

Thermal insulation

Stored

Kinetic energy

Dissipation

Transferred

Elastic potential energy

Changes in energy

Power

Specific heat capacity

ENERGY

Gravitational potential energy

Work

Global energy

Power

Renewable

Non-renewable

Practice questions

1. a) Explain the difference between a renewable and a non-renewable energy resource. **(2 marks)**

    b) Nuclear power is a very reliable source of energy but some environmentalists protest against it.
        Explain why. **(2 marks)**

    c) Many environmentalists would rather use wind or tidal power stations.
        What are the potential disadvantages of using these energy resources? **(2 marks)**

2. A car has a mass of 1600 kg and moves at 32 m/s.

    a) What is the kinetic energy of the car? **(2 marks)**

    b) If the fuel burnt by the car to reach this speed contained 1500 kJ, what was the efficiency of this transfer of energy? **(2 marks)**

    c) Give two ways energy was transferred into less useful forms by the car's engine. **(2 marks)**

    d) Give one feature of the car's engine that reduced the energy dissipated. **(1 mark)**

3. A bungee jumper of mass 77 kg stands on a bridge 150 m high.

    a) What is their gravitational potential energy? (Assume gravitational field strength = 10 N/kg) **(2 marks)**

    b) The bungee jumper jumps off and quickly reaches a speed of 20 m/s.
        What is their kinetic energy at this point? **(2 marks)**

    c) Explain the energy transfers that occur from when the bungee jumper jumps to when they reach their lowest height. **(2 marks)**

Transverse and longitudinal waves

Transverse waves

In a transverse wave, the oscillations are perpendicular to the direction of energy transfer, such as the ripples on the surface of water.

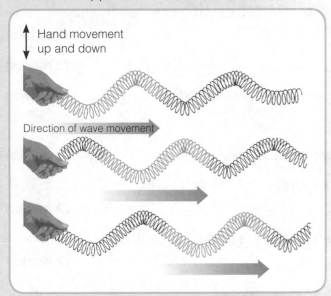

Hand movement up and down

Direction of wave movement

Longitudinal waves

In a longitudinal wave, the oscillations are parallel to the direction of energy transfer. Longitudinal waves show areas of **compression** and **rarefaction**, such as sound waves travelling through air.

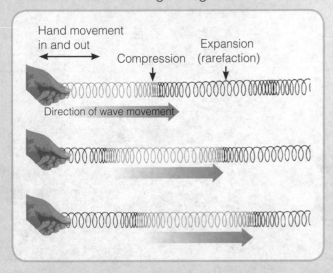

Hand movement in and out

Compression

Expansion (rarefaction)

Direction of wave movement

Wave movement

In both sound waves in air and ripples on the water surface, it is the wave that moves forward rather than the air or water molecules. The waves transfer energy and information without transferring matter.

This can be shown experimentally. For example, when a tuning fork is used to create a sound wave that moves out from the fork, the air particles don't move away from the fork. (This would create a vacuum around the tuning fork.)

Properties of waves

Waves are described by their:

➤ **amplitude** – The amplitude of a wave is the maximum displacement of a point on a wave away from its undisturbed position.

➤ **wavelength** – The wavelength of a wave is the distance from a point on one wave to the equivalent point on the adjacent wave.

➤ **frequency** – The frequency of a wave is the number of waves passing a point each second.

➤ **period** – The time for one complete wave to pass a fixed point. The equation for the time period (T) of a wave is given by the following equation:

$$\text{period} = \frac{1}{\text{frequency}}$$

$$\text{period} = \frac{1}{f}$$

➤ period, T, in seconds, s

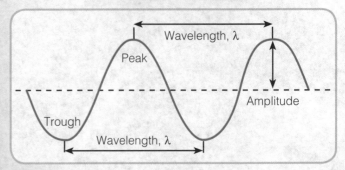

Wavelength, λ

Peak

Amplitude

Trough

Wavelength, λ

Keywords

Compression ➤ A region in a longitudinal wave where the particles are closer together

Rarefaction ➤ A region in a longitudinal wave where the particles are further apart

Wave speed

The wave speed is the speed at which the energy is transferred (or the wave moves) through the medium.

The wave speed is given by the following wave equation:

> **wave speed = frequency × wavelength**
>
> $$v = f\lambda$$
>
> ➤ wave speed, *v*, in metres per second, m/s
> ➤ frequency, *f*, in hertz, Hz
> ➤ wavelength, λ, in metres, m

Example:

A sound wave in air has a frequency of 250 Hz and a wavelength of 1.32 m. What is the speed of the sound wave?

wave speed = frequency × wavelength
= 250 × 1.32
= 330 m/s

A ripple tank and a stroboscope can be used to measure the speed of ripples on the surface of water.

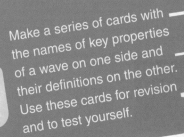

Make a series of cards with the names of key properties of a wave on one side and their definitions on the other. Use these cards for revision and to test yourself.

1. Give an example of a longitudinal wave and an example of a transverse wave.
2. Define the term 'wavelength'.
3. What is the speed of a wave with a frequency of 560 Hz and a wavelength of 3.2 m?

75

Electromagnetic waves and properties

Electromagnetic waves

Electromagnetic waves are transverse waves that:

➤ transfer energy from the source of the waves to an absorber
➤ form a continuous spectrum
➤ travel at the same velocity through a vacuum (space) or air.

The waves that form the electromagnetic spectrum are in terms of their wavelength and their frequency.

Make separate cards of the names of each electromagnetic radiation, its uses and any associated dangers. Mix up all these cards and then match them up.

The electromagnetic spectrum

Low frequency ←→ High frequency

| Radio waves | Micro-waves | Infrared radiation | Visible light | Ultraviolet | X-rays | Gamma rays |

10^3m   1 m 10^{-3}m 10^{-5}m 10^{-7}m 10^{-9}m 10^{-11}m 10^{-13}m
Wavelength

Visible light

Human eyes only detect visible light so only identify a limited range of electromagnetic radiation.

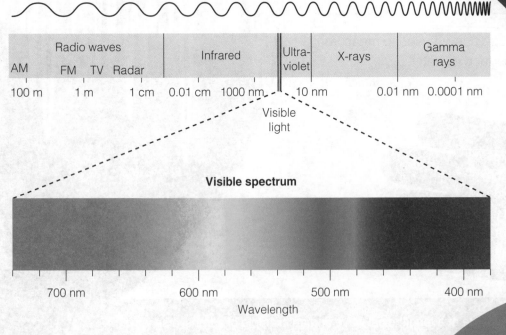

	Radio waves				Infrared		Ultra-violet	X-rays		Gamma rays
AM	FM	TV	Radar							
100 m	1 m		1 cm	0.01 cm	1000 nm		10 nm	0.01 nm		0.0001 nm

Visible light

Visible spectrum

700 nm   600 nm   500 nm   400 nm
Wavelength

Keywords

Gamma rays ➤ Electromagnetic radiation that originates from changes in the nucleus of a atom
Sievert ➤ Unit of radiation dose

Features of electromagnetic waves

Changes in atoms and the nuclei of atoms can result in electromagnetic waves being generated or absorbed over a wide frequency range.

Gamma rays originate from changes in the nucleus of an atom.

The diagram below shows the main features of electromagnetic waves.

Ultraviolet waves, x-rays and gamma rays can have hazardous effects on human body tissue.

The effects of electromagnetic waves depend on the type of radiation and the size of the dose.

1000 millisieverts (mSv) equals 1 sievert (Sv)

Radiation dose (in **sieverts**) is a measure of the risk of harm from an exposure of the body to the radiation.

Ultraviolet waves can cause skin to age prematurely and increase the risk of skin cancer.

X-rays and gamma rays are ionising radiation that can cause the mutation of genes and cancer.

Uses of electromagnetic waves

Electromagnetic waves have many practical applications.

Radio waves	Television, radio
Microwaves	Satellite communications, cooking food
Infrared	Electrical heaters, cooking food, infrared cameras
Visible light	Fibre optic communications
Ultraviolet	Energy efficient lamps, sun tanning
X-rays	Medical imaging and treatments
Gamma rays	Medical imaging and treatment of cancer

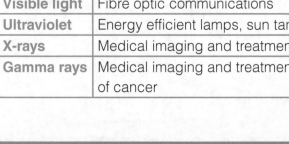

1. Which has the largest wavelength: infrared radiation or ultraviolet radiation?
2. What section of electromagnetic radiation can be perceived by the human eye?
3. What are the potential dangers of gamma rays?
4. Give one use of ultraviolet radiation.

Mind map

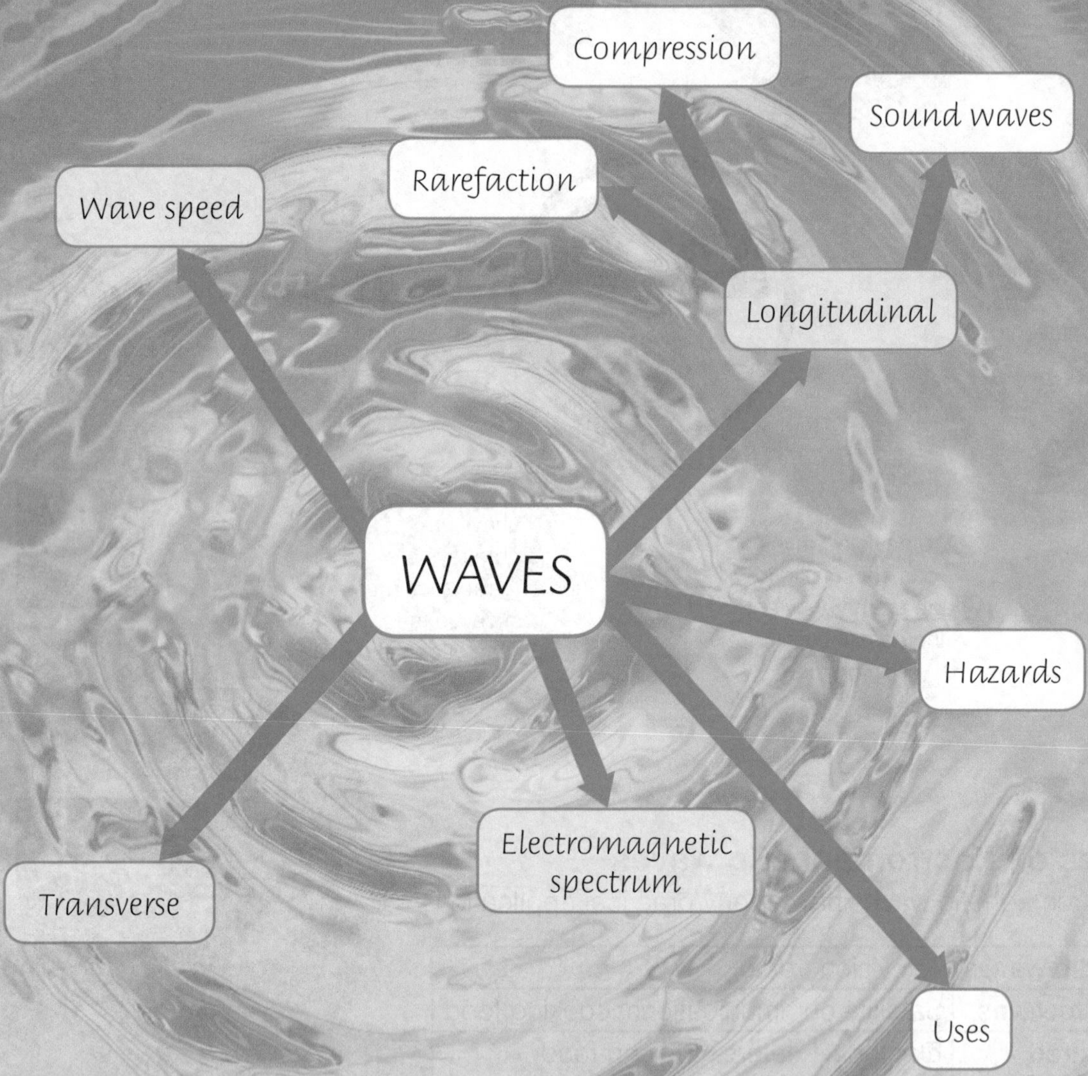

Compression

Sound waves

Rarefaction

Wave speed

Longitudinal

WAVES

Hazards

Transverse

Electromagnetic
spectrum

Uses

Practice questions

1. a) How does compression differ from rarefaction? **(2 marks)**

   b) What is the frequency of a wave travelling at 200 m/s with a wavelength of 3.2 m? **(2 marks)**

2. The following is an excerpt from a government health leaflet.

   When out in the sun during the summer, it's important to wear sun cream, particularly during the hottest parts of the day, to avoid 'sunburn'. Sunburn can lead to long term health problems.

   a) What type of radiation is this leaflet warning about? **(1 mark)**

   b) What long term health problems could this type of radiation lead to? **(1 mark)**

   c) Why can't this radiation be detected by the human eye? **(1 mark)**

3. David wrote the following explanation of how sound travels:

   'Sound travels as a transverse wave. When a sound wave travels in air the oscillations of the air particles are perpendicular to the direction of energy transfer. The sound wave transfers matter and energy.'

   Write this paragraph out again, correcting the three mistakes David has made. **(3 marks)**

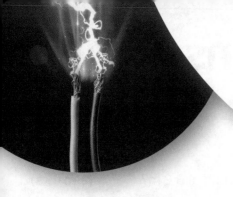

Circuits, charge and current

Circuit symbols

The diagram below shows the standard symbols used for components in a circuit.

Here is an example of a circuit diagram:

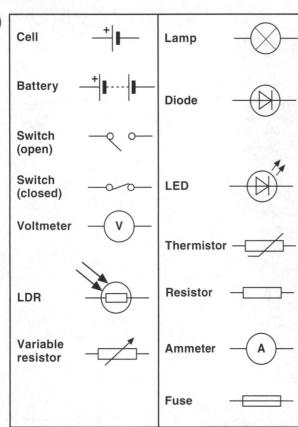

Cell		Lamp	
Battery		Diode	
Switch (open)			
Switch (closed)		LED	
Voltmeter		Thermistor	
LDR		Resistor	
Variable resistor		Ammeter	
		Fuse	

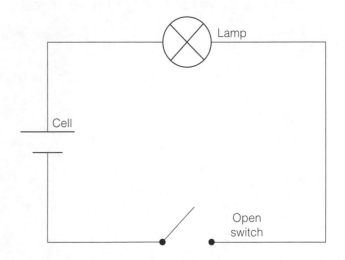

Lamp

Cell

Open switch

Resistors

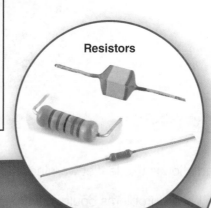

Fuses

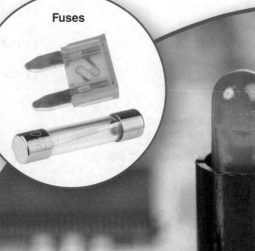

Diodes

Switches

AMPERES

Electrical charge and current

For electrical charge to flow through a closed circuit, the circuit must include a source of energy that produces a potential difference, such as a battery, cell or powerpack.

Electric current is a flow of electrical charge. The size of the electric current is the rate of flow of electrical charge. Charge flow, current and time are linked by the following equation:

> **charge flow = current × time**
>
> $Q = It$
> ➤ charge flow, Q, in coulombs, C
> ➤ current, I, in amperes or amps A
> ➤ time, t, in seconds, s

Example:

A current of 6 A flows through a circuit for 14 seconds. What is the charge flow?

charge flow = current × time
$$= 6 \times 14$$
$$= 84 \text{ C}$$

The current at any point in a single closed loop of a circuit has the same value as the current at any other point in the same closed loop.

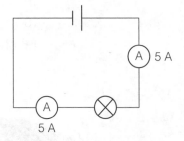

Both ammeters in the circuit above show the same current of 5 amps.

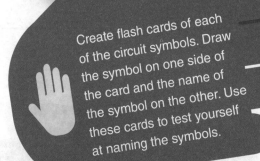

Create flash cards of each of the circuit symbols. Draw the symbol on one side of the card and the name of the symbol on the other. Use these cards to test yourself at naming the symbols.

1. What is the difference between the symbol for a resistor and the symbol for a variable resistor?
2. What charge flows through a circuit per second if the current is 4.2 A?
3. What is the current of a charge flow of 450 C in five seconds?

76

Current, resistance and potential difference

Current, resistance and potential difference

The current through a component depends on both the resistance of the component and the potential difference (p.d.) across the component.

The greater the resistance of the component, the smaller the current for a given potential difference across the component.

Current, potential difference or resistance can be calculated using the following equation:

> **potential difference = current × resistance**
>
> $$V = IR$$
>
> ➤ **potential difference, V, in volts, V**
> ➤ **current, I, in amperes or amps, A**
> ➤ **resistance, R, in ohms, Ω**

Example:

A 5 ohm resistor has a current of 2 A flowing through it. What is the potential difference across the resistor?

potential difference = current × resistance
$$= 5 \times 2$$
$$= 10 \text{ V}$$

By measuring the current through, and potential difference across a component, it's possible to calculate the resistance of a component.

The circuit diagram below would allow you to determine the resistance of the filament lamp.

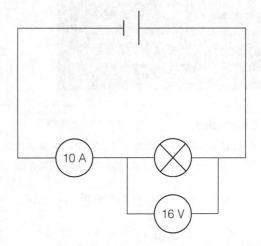

potential difference = current × resistance

$$\text{resistance} = \frac{\text{potential difference}}{\text{current}}$$

$$= \frac{16}{10}$$

$$= 1.6 \text{ ohms}$$

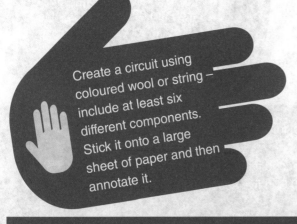

Create a circuit using coloured wool or string – include at least six different components. Stick it onto a large sheet of paper and then annotate it.

Keyword

Ohmic conductor ➤ Resistor at constant temperature where the current is directly proportional to the potential difference

Resistors

For some resistors, the value of R remains constant but in others (e.g. variable resistors) R can change as the current changes.

In an **ohmic conductor**, at a constant temperature the current is directly proportional to the potential difference across the resistor. This means that the resistance remains constant as the current changes.

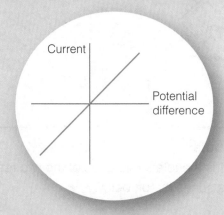

The resistance of components such as lamps, diodes, thermistors and LDRs is not constant; it changes with the current through the component. They are not ohmic conductors.

Filament lamps

The resistance of a filament lamp increases as the temperature of the filament increases.

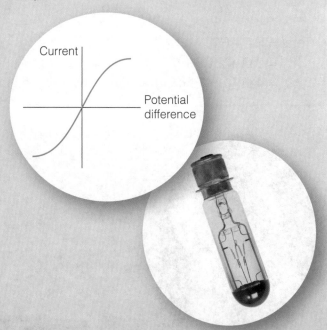

Diodes

The current through a diode flows in one direction only. This means the diode has a very high resistance in the reverse direction.

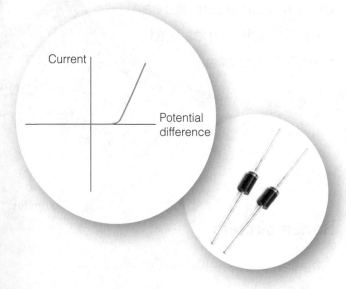

Light dependent resistors (LDR)

The resistance of an LDR decreases as light intensity increases. LDRs are used in circuits where lights are required to switch on when it gets dark, such as floodlights.

Thermistors

The resistance of a thermistor decreases as the temperature increases. Thermistors are used in thermostats to control heating systems.

1. Explain why an LDR is not an ohmic conductor.
2. What two pieces of equipment need to be wired into a circuit in order to determine the resistance of a component in the circuit?
3. What is the potential difference if the current is 4 A and the resistance is 2 ohms?
4. Calculate the resistance of a component that has a current of 3 A flowing through it and a potential difference of 6 V.

Components can be joined together in either a **series circuit** or a **parallel circuit**. Some circuits can include both series and parallel sections.

Series and parallel circuits

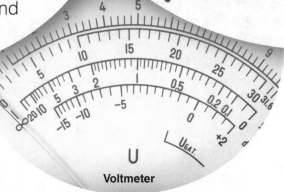

U

Voltmeter

Series circuits

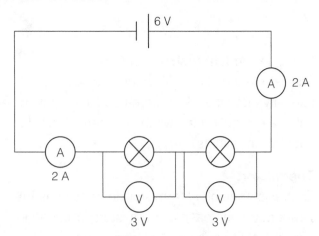

6 V

A) 2 A

A
2 A

V
3 V

V
3 V

For components connected in series:

➤ there is the same current through each component
➤ the total potential difference of the power supply is shared between the components
➤ the total resistance of two components is the sum of the resistance of each component.

Total resistance is given by the following equation:

$$R\,\text{total} = R1 + R2$$
➤ resistance, *R*, in ohms, Ω

Example:
What is the total resistance of the two resistors in the series circuit below?

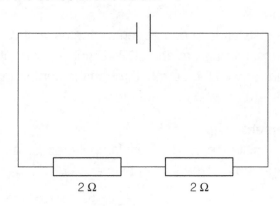

2 Ω 2 Ω

Total = R1 + R2
= 2 Ω + 2 Ω
= 4 Ω

Keywords

Series circuit ➤ Circuit where all components are connected along a single path
Parallel circuit ➤ Circuit which contains branches and where all the components will have the same voltage

Parallel circuits

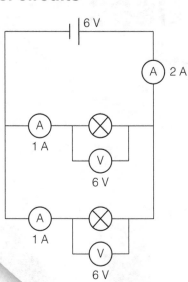

6 V

A 2 A

A 1 A

V 6 V

A 1 A

V 6 V

For components connected in parallel:

➤ the potential difference across each component is the same
➤ the total current through the whole circuit is the sum of the currents through the separate components. The current splits between the branches of the circuit and combines when the branches meet
➤ the total resistance of two resistors is less than the resistance of the smallest individual resistor. This is due to the potential difference across the resistors being the same but the current splitting.

78

Draw a parallel circuit of your own design, making sure that you include at least two resistors.

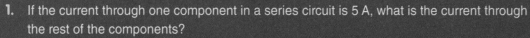

$I = \frac{V}{R}$

1. If the current through one component in a series circuit is 5 A, what is the current through the rest of the components?
2. How do you calculate the total resistance of two resistors in series?
3. A 3 Ω resistor and a 2 Ω resistor are connected in parallel. Would the resistance be higher or lower than 2 Ω?

Domestic uses and safety

Mains electricity

WS Most electrical appliances are connected to the mains using a three-core cable with a three-pin plug.

Live wire	Brown	Carries the alternating potential difference from the supply.
Neutral wire	Blue	Completes the circuit. The neutral wire is at, or close to, earth potential (0 V).
Earth wire	Green and yellow stripes	The earth wire is a safety wire to stop the appliance becoming live and is at 0 V. It only carries a current if there is a fault.

Direct and alternating potential difference

Cells and batteries supply current that always passes in the same direction. This is direct current (**dc**). This means the potential difference remains constant. For example, a 9 V battery supplies 9V continuously.

The current of mains electricity changes direction. This is alternating current (**ac**). Mains electricity in the UK has a frequency of 50 Hz – this means that the current changes direction fifty times per second. This causes the potential difference of a live mains wire to vary between a large positive voltage and a large negative voltage. The mean potential difference of the UK mains supply is about 230 V.

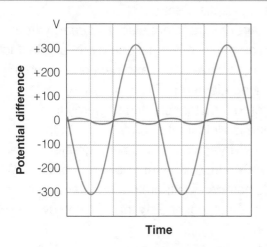

The graph shows the potential difference of two wires in a UK mains electricity cable. The red line shows the changes in potential difference in the live wire. The blue line shows the neutral wire, which has a potential difference of 0V. The mean potential difference is 230V but peaks and troughs of the live wire are greater than 230V.

The potential difference between the live wire and earth (0 V) is about 230 V.

Our bodies are at earth potential (0 V). Touching a live wire produces a large potential difference across our body. This causes a current to flow through our body, resulting in an electric shock that could cause serious injury or death.

1. What colour is the neutral wire?
2. What is the function of the earth wire?
3. Why is it dangerous to touch a live wire?

Keywords

dc ➤ Direct current that always passes in the same direction

ac ➤ Alternating current that changes direction

A live wire may be dangerous even when a switch in the mains circuit is open.

The fuse in a plug provides protection if a fault occurs. In the event of a fault, a large current flows from the live wire to earth. This melts the fuse and disconnects the live wire. Devices called circuit breakers work in a similar way. They respond more rapidly than a fuse to current changes – this makes them safer and they can be reset once the fault has been corrected.

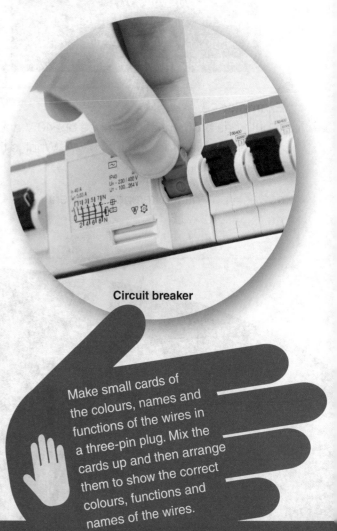

Circuit breaker

Make small cards of the colours, names and functions of the wires in a three-pin plug. Mix the cards up and then arrange them to show the correct colours, functions and names of the wires.

Power

The power transfer in any circuit device is related to the potential difference across it and the current through it, and to the energy changes over time. Power can be calculated using the following equations.

power = potential difference × current

$$P = VI$$

or

power = current² × resistance

$$P = I^2 R$$

➤ power, P, in watts, W
➤ potential difference, V, in volts, V
➤ current, I, in amperes or amps, A
➤ resistance, R, in ohms, Ω

Example:

A bulb has a potential difference of 240 V and a current flowing through it of 0.6 A. What is the power of the bulb?

power = potential difference × current
= 240 × 0.6
= 144 W

Energy transfers in everyday appliances

Everyday electrical appliances are designed to bring about energy transfers.

The amount of energy an appliance transfers depends on how long the appliance is switched on for and the power of the appliance.

Here are some examples of everyday energy transfer in appliances:

➤ A hairdryer transfers electrical energy from the ac mains to kinetic energy (in an electric motor to drive a fan) and heat energy (in a heating element).

➤ A torch transfers electrical energy from batteries into light energy from a bulb.

➤ A hair dryer will have a much greater power (2200W) than a battery powered torch (10W). The hair dryer is therefore transferring much more electrical energy than a torch over a given time.

Work done

Work is done when charge flows in a circuit.

The amount of energy transferred by electrical work can be calculated using the following equation:

> **energy transferred = power × time**
>
> $$E = Pt$$
>
> **and**
>
> **energy transferred** = **charge flow × potential difference**
>
> $$E = QV$$
>
> ➤ energy transferred, *E*, in joules, J
> ➤ power, *P*, in watts, W
> ➤ time, *t*, in seconds, s,
> ➤ charge flow, *Q*, in coulombs, C
> ➤ potential difference, *V*, in volts, V

The National Grid

The **National Grid** is a system of cables and transformers linking power stations to consumers.

Electrical power is transferred from power stations to consumers using the National Grid.

Step-up transformers **increase** the potential difference from the power station to the transmission cables.

Step-down transformers **decrease** the potential difference to a much lower and safer level for domestic use.

Increasing the potential difference reduces the current so reduces the energy loss due to heating in the transmission cables. Reducing the loss of energy through heat makes the transfer of energy much more efficient. Also, the wires would glow and be more likely to break over time if the current through them was high.

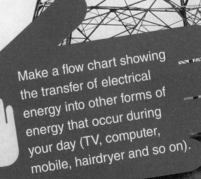

Make a flow chart showing the transfer of electrical energy into other forms of energy that occur during your day (TV, computer, mobile, hairdryer and so on).

1. What is the power of a device that has a current flowing through it of 5 A and a resistance of 2 Ω?
2. What is the energy transferred by a charge flow of 60 C and a potential difference of 12 V?
3. Why are step-down transformers important in the National Grid?

Magnetic fields and electromagnets

Keywords

Permanent magnet ➤ Magnet which produces its own magnetic field
Magnetic field ➤ Region around a magnet where a force acts on another magnet or on a magnetic material
Solenoid ➤ Coil wound into a helix shape

Poles of a magnet

The poles of a magnet are the places where the magnetic forces are strongest. When two magnets are brought close together they exert a force on each other.

> Two like poles repel. Two unlike poles attract.

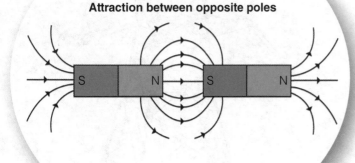

Attraction between opposite poles

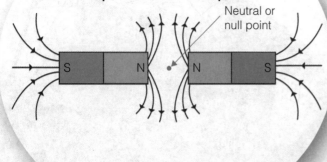

Repulsion between like poles

Neutral or null point

Magnetism is an example of a non-contact force.

Permanent magnetism vs induced magnetism

A **permanent magnet** . . .

➤ produces its own **magnetic field**.

An induced magnet . . .

➤ becomes a magnet when placed in a magnetic field
➤ always causes a force of attraction
➤ loses most or all of its magnetism quickly when removed from a magnetic field.

Magnetic field

The region around a magnet – where a force acts on another magnet or on a magnetic material (iron, steel, cobalt, magnadur and nickel) – is called the magnetic field.

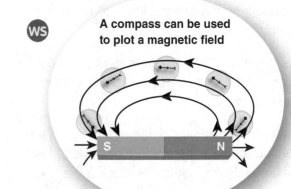

A compass can be used to plot a magnetic field

The **force** between a magnet and a magnetic material is always attraction.

The **strength** of the magnetic field depends on the distance from the magnet.

The **field** is strongest at the poles of the magnet.

The **direction** of a magnetic field line is from the north pole of the magnet to the south pole of the magnet.

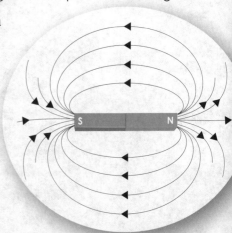

Compasses

A magnetic compass contains a bar magnet that points towards magnetic north. This provides evidence that the Earth's core is magnetic and produces a magnetic field.

Electromagnets

When a current flows through a conducting wire a magnetic field is produced around the wire.

The shape of the magnetic field can be seen as a series of concentric circles in a plane, perpendicular to the wire.

The direction of these field lines depends on the direction of the current.

The strength of the magnetic field depends on the current through the wire and the distance from the wire.

Coiling the wire into a **solenoid** (a helix) increases the strength of the magnetic field created by a current through the wire.

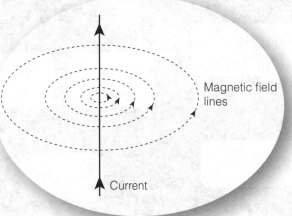

Magnetic field lines

Current

Features of a solenoid

Adding an iron core increases the magnetic field strength of a solenoid.

Magnetic field has a similar shape to that of a bar magnet.

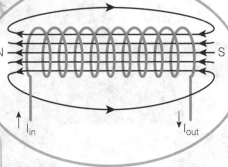

N

S

I_{in}

I_{out}

The fields from individual coils in the solenoid add together to form a very strong, almost uniform, field along the centre of the solenoid.

An electromagnet is a solenoid with an iron core.

The fields cancel to give a weaker field outside the solenoid.

Make your own compass using a bowl of water, a small dish, a magnet and a nail. To make a magnet for your compass, take the nail and magnet and stroke one end of the magnet along the length of the nail, always going in the same direction. Once your nail is magnetised, you need to float it in the water. Start by putting a dish into the water, then place the nail into the centre of the dish. Let it settle a bit, to move around a little bit, until it has found north.

1. What happens if the two north poles of a bar magnet are brought together?
2. When will a magnetic material become an induced magnet?
3. Where is the magnetic field of a magnet strongest?
4. Why does a magnetic compass point north?
5. What two things does the strength of a magnetic field around a wire depend on?
6. What shape is the magnetic field around a solenoid?

Mind map

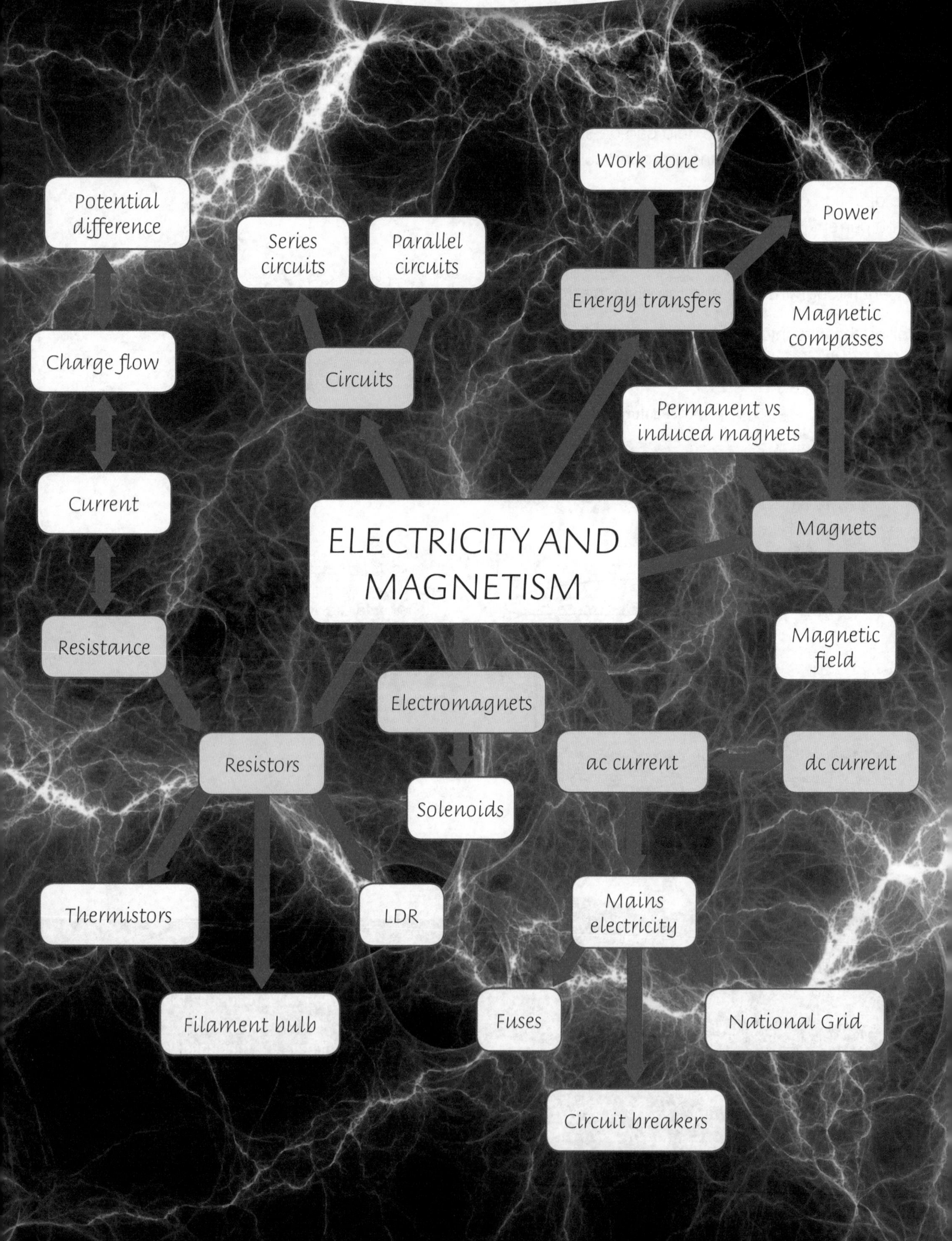

Work done

Potential difference

Series circuits

Parallel circuits

Power

Energy transfers

Magnetic compasses

Charge flow

Circuits

Permanent vs induced magnets

Current

ELECTRICITY AND MAGNETISM

Magnets

Resistance

Magnetic field

Electromagnets

Resistors

ac current

dc current

Solenoids

Thermistors

LDR

Mains electricity

Filament bulb

Fuses

National Grid

Circuit breakers

Practice questions

1. Look at the parallel circuit on the right.

   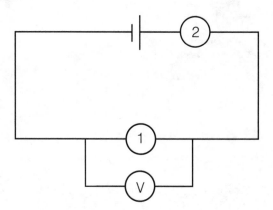

   a) Component 1 has a resistance of 3 Ω and has a potential difference of 15 V across it. What is the current through it? **(3 marks)**

   b) What is the current at point 2 in the circuit? Explain how you arrived at your answer. **(2 marks)**

   c) Sketch a graph to show the relationship between potential difference and current in an ohmic conductor. **(3 marks)**

   d) Give an example of a component that would show a non-linear relationship between potential difference and current. **(1 mark)**

2. a) What colour is the live wire in an electrical plug? **(1 mark)**

   b) Explain the importance of this wire being a different colour to the neutral wire. **(2 marks)**

3. The diagram below shows a bar magnet.

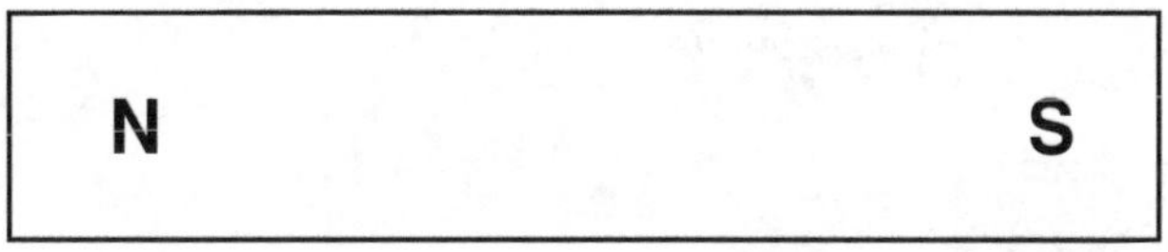

N   S

   a) Draw the shape of the magnetic field around this magnet. **(2 marks)**

   b) What would happen if the south pole of a bar magnet was brought into contact with the north pole of another magnet? **(1 mark)**

   c) Give two differences between a bar magnet and an induced magnet. **(2 marks)**

4. A wire is used in an investigation into magnetism.

   a) How can a magnetic field be produced around a wire? **(1 mark)**

   b) What effect would coiling the wire into a solenoid have? **(1 mark)**

   c) How could this wire be turned into an electromagnet? **(1 mark)**

The particle model and pressure

The particle model

Matter can exist as a solid, liquid or as a gas.

Solid – particles are very close together and vibrating. They are in fixed positions.

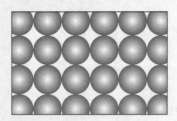

Liquid – particles are very close together but are free to move relative to each other. This allows liquids to flow.

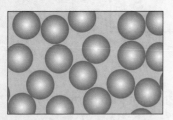

Gas – particles in a gas are not close together. The particles move rapidly in all directions.

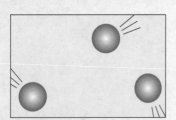

If the particles in a substance are more closely packed together, the density of the substance is higher. This means that liquids have a higher density than gases. Most solids have a higher density than liquids.

Density also increases when the particles are forced into a smaller volume.

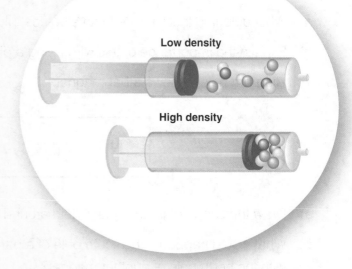

Low density

High density

The density of a material is defined by the following equation:

$$\text{density} = \frac{\text{mass}}{\text{volume}}$$

$$\rho = \frac{m}{V}$$

➤ density, ρ, in kilograms per metre cubed, kg/m³
➤ mass, m, in kilograms, kg
➤ volume, V, in metres cubed, m³

Example:

What is the density of an object that has a mass of 56 kg and a volume of 0.5 m³?

$$\rho = \frac{m}{V}$$

$$= \frac{56}{0.5} = 112 \text{ kg/m}^3$$

When substances change state (melt, freeze, boil, evaporate, condense or sublimate), mass is conserved (it stays the same).

Changes of state are physical changes: the change does not produce a new substance, so if the change is reversed the substance recovers its original properties.

Ice

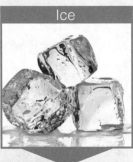

Water

Steam

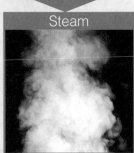

Gas under pressure

The molecules of a gas are in constant random motion.

When the molecules collide with the wall of their container they exert a force on the wall. The total force exerted by all of the molecules inside the container on a unit area of the wall is the **gas pressure**.

Increasing the temperature of a gas, held at constant volume, **increases** the pressure exerted by the gas.

Decreasing the temperature of a gas, held at constant volume, **decreases** the pressure exerted by the gas.

The temperature of the gas is related to the average kinetic energy of the molecules. The higher the temperature, the greater the average kinetic energy, and so the faster the average speed of the molecules. At higher temperatures, the particles collide with the walls of the container at a higher speed.

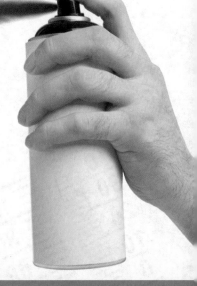

Draw a comic strip showing the effect of changing volume and temperature on a gas. Annotate your comic strip using the terms and equations in this module.

1. Explain how gases exert pressure.
2. What is the density of a 2 kg object with a volume of 0.002 m³?
3. What happens to the mass of a liquid when it freezes?

Internal energy and change of state

Internal energy

Energy is stored inside a system by the particles (atoms and molecules) that make up the system. This is called **internal energy**.

Internal energy of a system is equal to the total kinetic energy and potential energy of all the atoms and molecules that make up the system.

Heating changes the energy stored within the system by increasing the energy of the particles that make up the system. This either raises the temperature of the system or produces a change of state.

Changes of heat and specific latent heat

When a change of state occurs, the stored internal energy changes, but the temperature remains constant. The graph below shows the change in temperature of water as it is heated; the temperature is constant when the water is changing state.

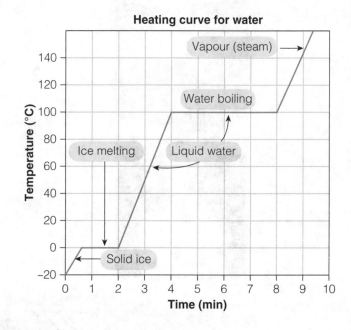

Heating curve for water

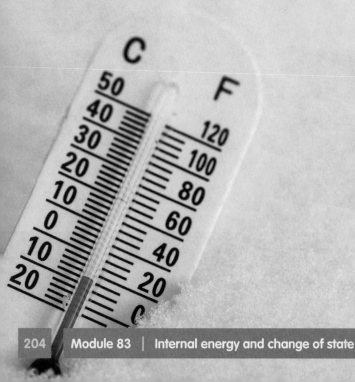

The **specific latent heat** of a substance is equal to the energy required to change the state of one kilogram of the substance with no change in temperature.

Keyword

Specific latent heat ➤ The energy required to change the state of one kilogram of the substance with no change in temperature

The energy required to cause a change of state can be calculated by the following equation:

> **energy for a change of state** = mass × specific latent heat
>
> $$E = mL$$
>
> ➤ energy, E, in joules , J
> ➤ mass, m, in kilograms, kg
> ➤ specific latent heat, L, in joules per kilogram, J/kg

Example:
What is the energy needed for 600 g of water to melt? (The specific latent heat of water melting is 334 kJ/kg.)

$$E = mL$$
$$0.6 \times 334 = 200.4 \text{ kJ}$$

The specific latent heat of fusion is the energy required for a change of state from solid to liquid.

The specific latent heat of vapourisation is the energy required for a change of state from liquid to vapour.

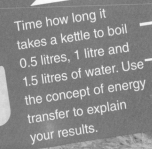

Time how long it takes a kettle to boil 0.5 litres, 1 litre and 1.5 litres of water. Use the concept of energy transfer to explain your results.

1. What energy is needed for 200 kg of cast iron to turn from a solid to a liquid? (The specific latent heat of iron melting is 126 kJ/kg.)
2. Explain the difference between the specific latent heat of fusion and the specific latent heat of vapourisation.
3. If a substance is heated but its temperature remains constant, what is happening to the substance?

Mind map

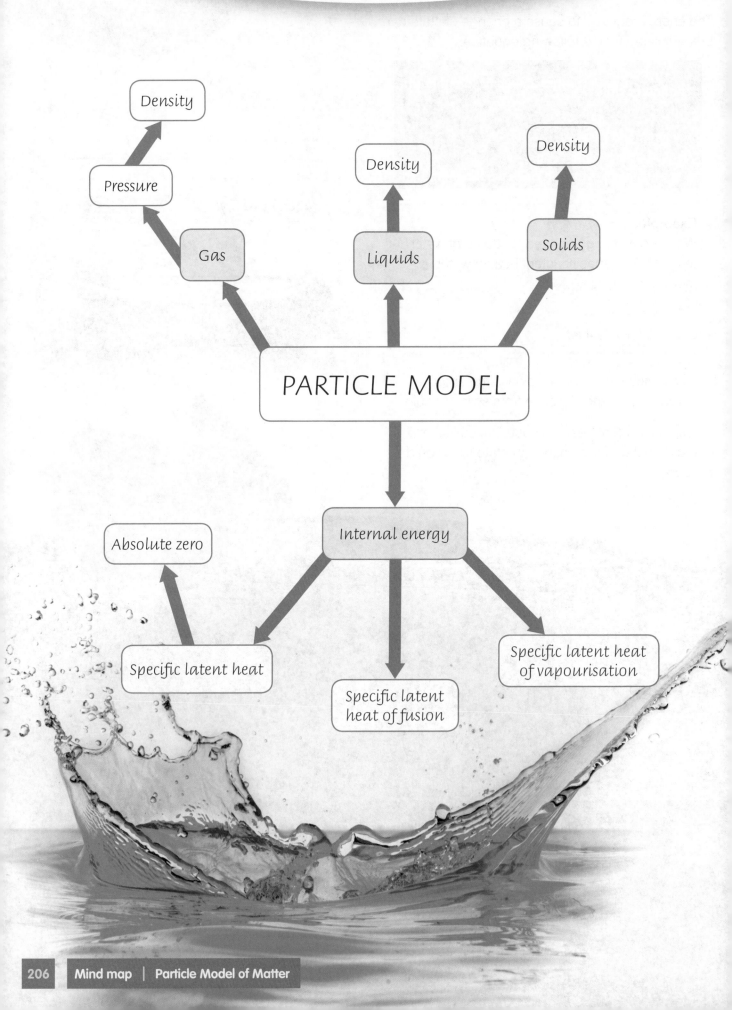

Density

Pressure

Gas

Density

Liquids

Density

Solids

PARTICLE MODEL

Internal energy

Absolute zero

Specific latent heat

Specific latent heat of fusion

Specific latent heat of vapourisation

Practice questions

1. A solid has a mass of 150 g and a volume of 0.0001 m³.

   a) What is the density of the solid? **(2 marks)**

   b) What happens to the mass and density of the solid when it sublimes? **(2 marks)**

   c) Explain what would happen to the stored internal energy and the temperature of the solid as it sublimed. **(2 marks)**

   d) The specific latent heat of the solid is 574 kJ/kg.

   What is the energy required for the solid to sublime? **(2 marks)**

2. Explain, using the particle model, why:

   a) a gas at a constant volume has a higher pressure when the temperature increases. **(2 marks)**

   b) a gas at a constant temperature has a lower pressure when the volume increases. **(2 marks)**

3. What term is given to the energy required for 1 kg of a liquid to become a gas? **(1 mark)**

Atoms and isotopes

Structure of atoms

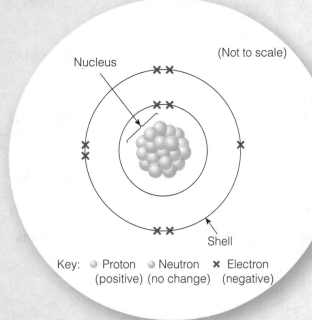

(Not to scale)

Nucleus

Shell

Key: ● Proton (positive) ● Neutron (no change) ✕ Electron (negative)

Atoms have a radius of around 1×10^{-10} metres.

The radius of a nucleus is less than $\frac{1}{10\,000}$ of the radius of an atom.

Most of the mass of an atom is concentrated in the nucleus. Protons and neutrons have a relative mass of 1 while electrons have a relative mass of 0.0005. The electrons are arranged at different distances from the nucleus (are at different energy levels).

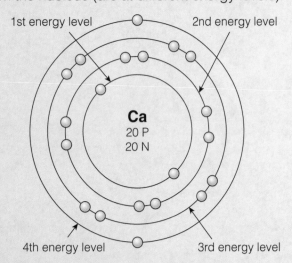

1st energy level

2nd energy level

Ca
20 P
20 N

4th energy level

3rd energy level

Absorption of electromagnetic radiation causes the electrons to become excited and move to a higher energy level and further from the nucleus.

Emission of electromagnetic radiation causes the electrons to move to a lower energy level and move closer to the nucleus.

If an atom loses or gains an electron, it is ionised.

The number of electrons is equal to the number of protons in the nucleus of an atom.

Atoms have no overall electrical charge.

All atoms of a particular element have the same number of protons. The number of protons in an atom of an element is called the **atomic number**.

The total number of protons and neutrons in an atom is called the **mass number**.

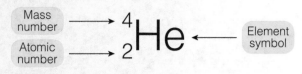

Mass number

Atomic number

$^{4}_{2}\text{He}$

Element symbol

Atoms of the same element can have different numbers of neutrons; these atoms are called isotopes of that element. For example, below are some isotopes of nitrogen. They each have 7 protons in the nucleus but different numbers of neutrons, giving the different isotopes.

$$^{14}\text{N} \quad ^{15}\text{N} \quad ^{13}\text{N}$$

Atoms turn into **positive ions** if they lose one or more outer electrons and into **negative ions** if they gain one or more outer electrons.

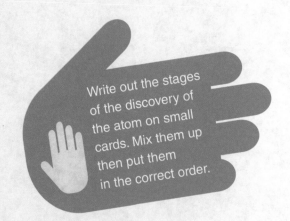

Write out the stages of the discovery of the atom on small cards. Mix them up then put them in the correct order.

The development of the atomic model

VS

Before the discovery of the electron, atoms were thought to be tiny spheres that could not be divided.

The discovery of the electron led to further developments of the model. The plum pudding model suggested that the atom is a ball of positive charge with negative electrons embedded in it.

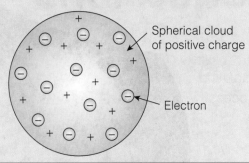

Spherical cloud of positive charge

Electron

Rutherford, Geiger and Marsden's alpha scattering experiment led to the conclusion that the mass of an atom was concentrated at the centre (nucleus) and that the nucleus was charged.

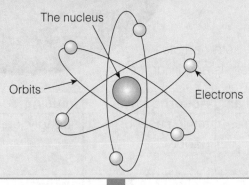

The nucleus

Orbits

Electrons

This evidence led to the nuclear model replacing the plum pudding model.

Niels Bohr suggested that the electrons orbit the nucleus at specific distances. The theoretical calculations of Bohr agreed with experimental observation.

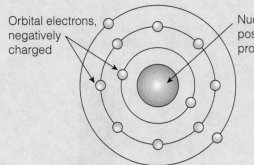

Orbital electrons, negatively charged

Nucleus, containing positively charged protons

Later experiments led to the idea of the nucleus containing smaller particles with the same amount of positive charge (**protons**).

The experimental work of James Chadwick provided the evidence to show the existence of neutrons within the nucleus. This was about 20 years after the nucleus became an accepted scientific idea.

Keywords

Atomic number ➤ Number of protons in an atom

Mass number ➤ Total number of protons and neutrons in an atom

1. What is the effect on the electrons when an atom absorbs electromagnetic radiation?
2. What name is given to atoms of the same element that have different numbers of neutrons?
3. What is the difference between the plum pudding model and the nuclear model?

Radioactive decay, nuclear radiation and nuclear equations

Radioactive decay

Some atomic nuclei are unstable. The nucleus gives out radiation as it changes to become more stable. This is a random process called **radioactive decay**.

Activity is the rate at which a source of unstable nuclei decays, measured in **becquerel** (Bq).

➤ 1 becquerel = 1 decay per second

Count rate is the number of decays recorded each second by a detector, such as a Geiger-Müller tube.

➤ 1 becquerel = 1 count per second

Radioactive decay can release a neutron, alpha particles, beta particles or gamma rays. If radiation is ionising, it can damage materials and living cells.

Particle	Description	Penetration in air	Absorbed by...	Ionising power
Alpha particles (α)	Two neutrons and two protons (a helium nucleus).	a few centimetres	a thin sheet of paper	strongly ionising
Beta particles (β)	High speed electron ejected from the nucleus as a neutron turns into a proton.	a few metres	a sheet of aluminium about 5 mm thick	moderately ionising
Gamma rays (γ)	Electromagnetic radiation from the nucleus.	a large distance	a thick sheet of lead or several metres of concrete	weakly ionising

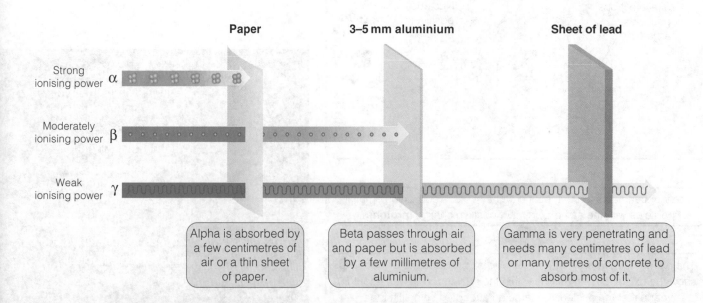

	Paper	**3–5 mm aluminium**	**Sheet of lead**

Strong ionising power α

Moderately ionising power β

Weak ionising power γ

Alpha is absorbed by a few centimetres of air or a thin sheet of paper.

Beta passes through air and paper but is absorbed by a few millimetres of aluminium.

Gamma is very penetrating and needs many centimetres of lead or many metres of concrete to absorb most of it.

Nuclear equations

Nuclear equations are used to represent radioactive decay.

Nuclear equations can use the following symbols:

$$^{4}_{2}\text{He}$$ alpha particle

$$^{0}_{-1}\text{e}$$ beta particle

Alpha decay causes both the mass and charge of the nucleus to decrease, as two protons and two neutrons are released.

$$^{219}_{86}\text{radon} \rightarrow {}^{215}_{84}\text{polonium} + {}^{4}_{2}\text{He}$$

Beta decay does not cause the mass of the nucleus to change but does cause the charge of the nucleus to increase, as a neutron becomes a proton.

$$^{14}_{6}\text{carbon} \rightarrow {}^{14}_{7}\text{nitrogen} + {}^{0}_{-1}\text{e}$$

The emission of a gamma ray does not cause the mass or the charge of the nucleus to change.

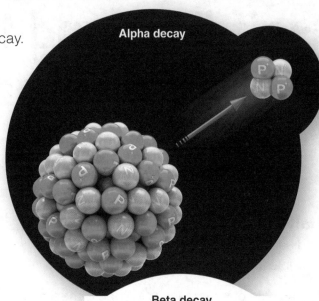

Alpha decay

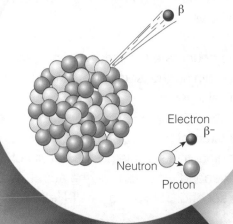

Beta decay

Electron

β⁻

Neutron
Proton

Make a large version of the radioactive decay table on the opposite page. Cut out each individual box to form separate cards. Mix the cards up and then group them to show the features of alpha, beta and gamma particles.

Keywords

Radioactive decay ➤ Random release of radiation from an unstable nucleus as it becomes more stable

Becquerel ➤ Unit of rate of radioactive decay

1. How far does alpha radiation penetrate in air?
2. What material is required to absorb gamma rays?
3. What effect does beta decay have on the mass and charge of the nucleus of an atom?

Half-lives and the random nature of radioactive decay

Uranium

Half-life

Radioactive decay occurs randomly. It is not possible to predict which nuclei will decay.

The half-life of a radioactive isotope is the time it takes for:

➤ the number of nuclei in a sample of the isotope to halve

or

➤ the count rate (or activity) from a sample containing the isotope to fall to half of its initial level.

> **For example:**
>
> A radioactive sample has an activity of 560 counts per second. After 8 days, the activity is 280 counts per second. This gives a half-life of 8 days.
>
> The decay can be plotted on a graph and the half-life determined from the graph.
>
>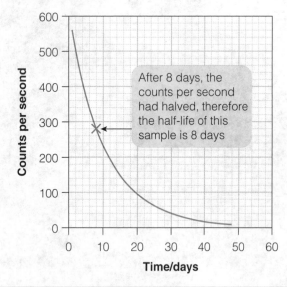
>
> After 8 days, the counts per second had halved, therefore the half-life of this sample is 8 days

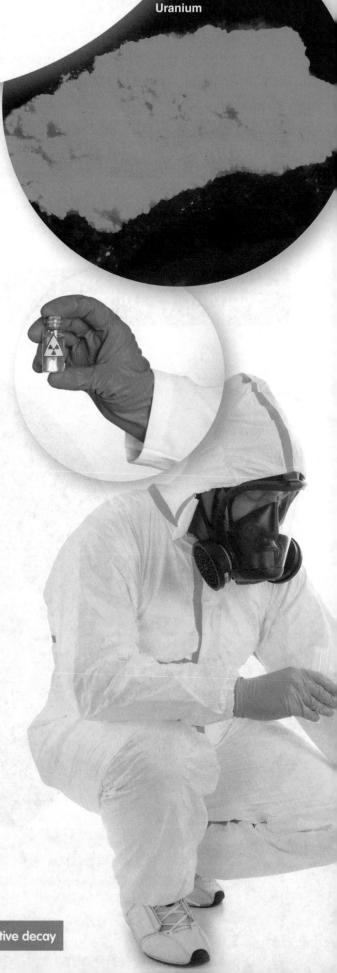

Radioactive contamination

Radioactive contamination is the unwanted presence of materials containing radioactive atoms or other materials.

This is a hazard due to the decay of the contaminating atoms. The level of the hazard depends on the type of radiation emitted.

Irradiation is the process of exposing an object to nuclear radiation. This is different from radioactive contamination as the irradiated object does not become radioactive.

Suitable precautions must be taken to protect against any hazard from the radioactive source used in the process of irradiation. In medical testing using radioactive sources, the doses patients receive are limited and medical staff wear protective equipment.

Keyword

Irradiation ➤ Exposing an object to nuclear radiation without the object becoming radioactive itself

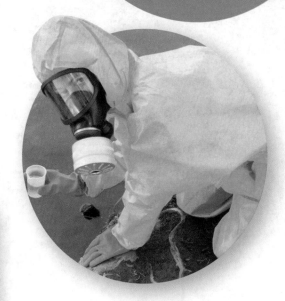

It is important for the findings of studies into the effects of radiation on humans to be published and shared with other scientists. This allows the findings of the studies to be checked by other scientists by the peer review process.

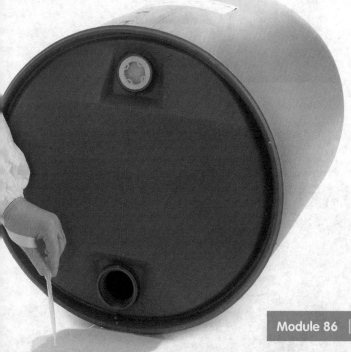

Shake ten or more coins in your hand and then drop them. Remove all the coins that are heads and record the number you've removed. Repeat the experiment until you have one or no coins left. Now plot a graph of how the number of coins in your hand decreased over time. Then, research graphs showing radioactive decay and compare them to the graph from your experiment.

1. Give the two definitions of half-life.
2. What is the half-life of a sample that goes from a count per second of 960 to 240 in 10 months?
3. Why is radioactive contamination a hazard?

Mind map

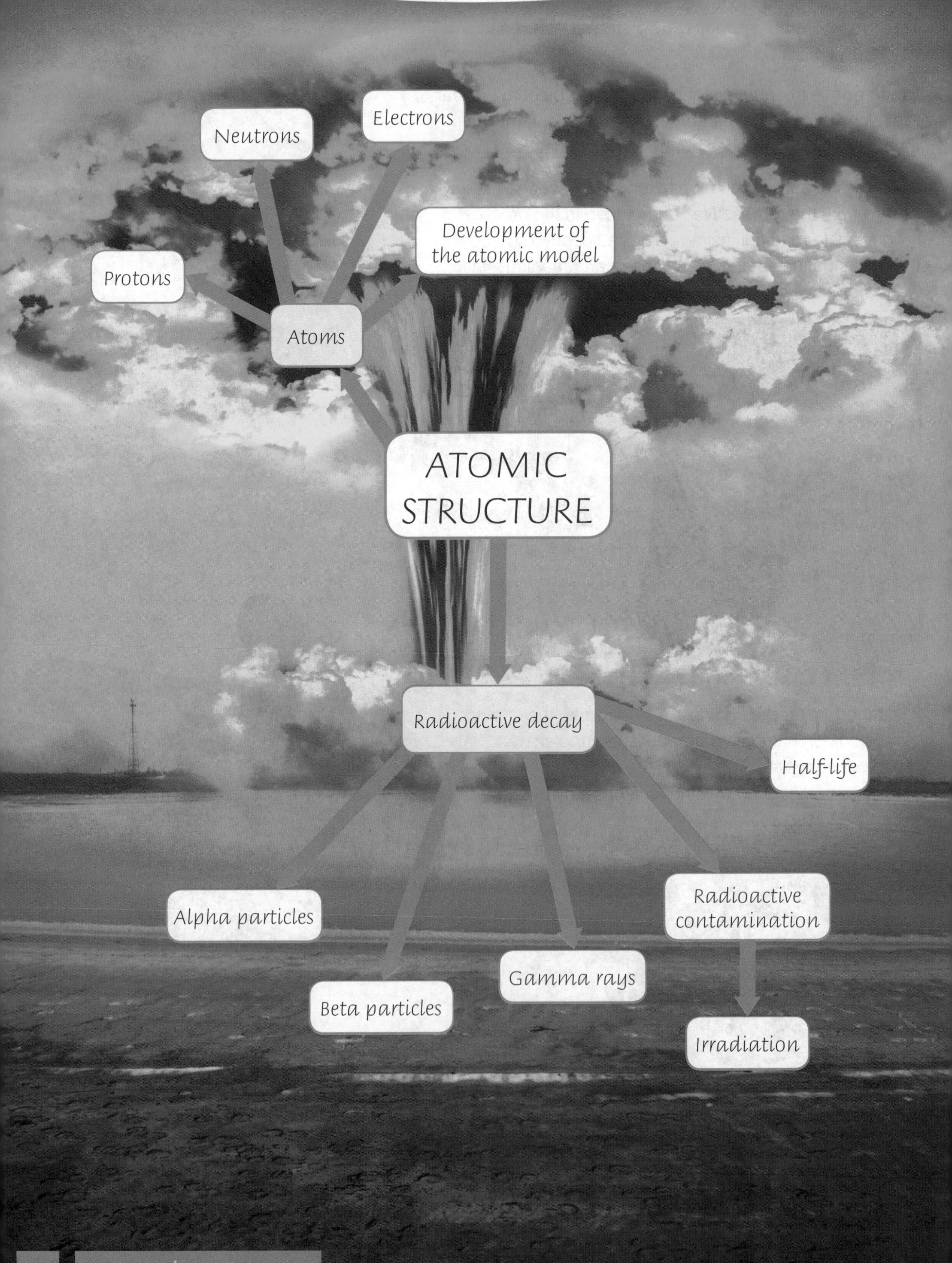

Neutrons

Electrons

Development of
the atomic model

Protons

Atoms

ATOMIC STRUCTURE

Radioactive decay

Half-life

Alpha particles

Radioactive
contamination

Beta particles

Gamma rays

Irradiation

Practice questions

1. The symbol below shows an element.

   $^{48}_{22}\text{Ti}$

   a) What is its . . .

      i)   number of protons? **(1 mark)**

      ii) number of electrons? **(1 mark)**

      iii) number of neutrons? **(1 mark)**

   b) For parts **i)–iii)** above, explain how you arrived at your answers. **(6 marks)**

2. Experiments by Rutherford, Geiger and Marsden led to the plum pudding model being replaced by the nuclear model.

   a) Explain the differences between the plum pudding model and the nuclear model. **(2 marks)**

   b) How did the further work of Bohr and the further work of Chadwick refine the nuclear model? **(2 marks)**

3. The graph below shows the activity of a radioactive sample over time.

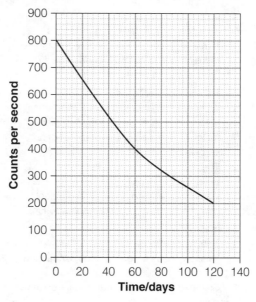

   a) What is the half-life of this sample?
      Explain how you arrived at your answer. **(2 marks)**

   b) Would this sample be useful as a radioactive tracer in medicine?
      Explain your answer. **(2 marks)**

Answers

Biology

Page 7

1. **Prokaryotes:** any three bacteria or archaebacteria, e.g. cholera, *E. coli* and salmonella. **Eukaryotes:** any three from the plant, animal, protist or fungal kingdoms, e.g. geranium (plant), tiger (animal), amoeba (protist) and mushroom (fungus).
2. A prokaryote has a DNA loop and plasmids. A eukaryote's DNA is found in the nucleus.
3. True – all cells do have a cell membrane.
4. Mitochondrion.
5. Chloroplasts contain chlorophyll to absorb sunlight for photosynthesis.

Page 9

1. An organ.
2. It is able to contract (shorten).
3. Phloem.
4. **Any one benefit**, e.g. can be used to treat serious conditions, cancer research, organ transplants. Any one objection, e.g. the embryo has the potential to be a living human and shouldn't be experimented with, risk of viral transmission.

Page 11

1. **Any two from:** electron microscopes use electrons to form images, light microscopes use light waves; electron microscopes produce 2D and 3D images, light microscopes use 2D images only; electron microscopes have a magnification up to ×500 000 (2D), light microscopes have a magnification of up to ×1500; electron microscopes are able to observe small organelles, light microscopes are only able to observe cells and larger organelles; electron microscopes enable you to see things at high resolution, light microscopes allow you to see things at low resolution.
2. The resolving power of a light microscope can only be used for objects that are greater than 200 nm across.

Page 13

1. For growth, repair and reproduction.
2. Chromosomes.
3. **Mitosis:** asexual reproduction, repair and growth; **meiosis:** sexual reproduction.
4. DNA/chromosomes and all other organelles.

Page 15

1. **Building** – any one from: converting glucose to starch in plants/glucose to glycogen in animals; synthesis of lipid molecules, formation of amino acids in plants (which are built up into proteins). **Breaking down** – any one from: breaking down excess proteins to form urea, respiration.
2. Aerobic.

3. **Any two from:** transmitting nerve impulses, active transport, muscle contraction/movement, maintaining a constant body temperature, synthesis of molecules.
4. **Humans:** lactic acid; **yeast:** ethanol and carbon dioxide.
5. To pay back the oxygen debt, i.e. take in oxygen to remove the lactic acid in muscles.

Page 17

1. pH and temperature.
2. A substrate molecule fits exactly into a specific enzyme's active site.
3. It emulsifies fat and helps neutralise acid from the stomach.
4. Amino acids.

Page 19

1. **A:** cell membrane, **B:** DNA, **C:** cell wall, **D:** plasmid. (**1 mark for each correct**)
2. a) Lactic acid increase in muscles causes muscle fatigue/tiredness (**1 mark**); lactic acid is toxic (**1 mark**).
   b) Anaerobic respiration releases lower amounts of energy than aerobic respiration; not enough energy available for extended periods of intense activity. (**1 mark**)
   c) Glucose is completely broken down (to carbon dioxide and water). (**1 mark**)
3. a) Denaturing/denaturation. (**1 mark**)
   b) Change in shape of active site (**1 mark**); substrate no longer fits active site (**1 mark**).
4. A specialised system is needed to transport oxygen over long distances (**1 mark**); paramecium can rely on simple diffusion due to its larger surface area / volume ratio (**1 mark**).

Page 21

1. A shrew – the proportion of its area to its volume is greater, despite the fact that its **total** body area is less than that of an elephant.
2. The plant cells have a higher water potential/lower solute concentration. The water moves **down** an osmotic gradient from inside the cells into the salt solution.

Page 23

1. Root hair cells.
2. To allow space for gases to be exchanged more freely.
3. So that cells can freely carry water up the plant.

Page 25

1. It would increase the (evapo)transpiration rate.
2. When temperatures are high (during the day) and loss of water through open stomata would cause the plant to dehydrate and wilt.
3. Osmosis.
4. Water enters the xylem at root level due to a 'suction' force caused by evaporation from the leaves, called (evapo)transpiration. The water is drawn up as a column within the stem's xylem.

Page 27

1. Valves prevent backflow of blood and ensure it reaches the heart (especially from parts of the body vertically below the heart).
2. Excess fat (particularly saturated fat) in the diet causes cholesterol to build up and block the coronary artery. This restricts the supply of blood to the heart muscle, which then does not receive enough oxygen and glucose. The heart muscle therefore dies.

Page 29

1. **Red blood cells** do not have a nucleus, are biconcave in shape, contain haemoglobin and carry oxygen. **Lymphocytes** have a nucleus, are irregular in shape and are involved in the immune response.
2. Oxygen is breathed into alveoli, diffuses across alveolar wall and combines with haemoglobin in a red blood cell. Blood enters the heart, from which it is pumped to the respiring muscle tissue. The oxygen diffuses from the red blood cell into a muscle cell, where it reacts with glucose during aerobic respiration.

Page 31

1. Photosynthesis takes **in** carbon dioxide and water, and requires an energy input. It produces glucose and oxygen. Respiration **produces** carbon dioxide and water, and releases energy. It absorbs glucose and oxygen.
2. As carbon dioxide concentration increases, so does the rate of photosynthesis. However, a maximum rate is eventually reached where no further increase in photosynthesis rate occurs despite increases in carbon dioxide concentration.

Page 33

1. **a**, **d** and **e** (**1 mark** for each correct answer)
2. a) Healthy man (**1 mark**)
   b) **Any two from:** less oxygen absorbed into blood; longer diffusion distance across alveolar wall; insufficient oxygen delivered to muscles to release energy for exercise (**2 marks**)
3. a) **A:** artery (**1 mark**); **B:** vein (**1 mark**).
   b) **Any one from:** an artery has to withstand/recoil with higher pressure; elasticity allows smoother blood flow/second boost to blood when recoils. (**1 mark**)
   c) **Any one from:** to prevent backflow of blood; compensate for low blood pressure. (**1 mark**)

Page 35

1. **Any three from:** smoking tobacco, drinking excess alcohol, carcinogens, ionising radiation.
2. The body is more prone to infections.

Page 37

1. More establishments would continue to sell infected food, therefore the number of salmonella cases would increase.
2. HIV requires use of retroviral drugs; Gonorrhoea can be treated by administering antibiotic injections followed by antibiotic tablets.
3. Photosynthesis and growth is affected.

Page 39

1. Phagocyte engulfs a pathogen where it is digested by the cell's enzymes.
2. Lymphocytes detect antigen; this triggers production of antibodies; once pathogens are destroyed, memory cells are produced; further infection with same pathogen dealt with swiftly, as antibodies produced rapidly in large numbers. The process can also be triggered via vaccination.

Page 41

1. An antibiotic is a drug that kills bacteria; an antibody is a protein produced by the immune system that kills bacteria.
2. An antiviral alleviates symptoms of an infection; analgesics are painkillers.
3. Do not prescribe antibiotics for viral/non-serious infections; ensure that the full course of antibiotics is completed.

Page 43

1. Human volunteers; computer simulations/modelling; cells grown in tissue culture.
2. **Any two from:** to see if they work; to make sure they are safe; to make sure that they are given at the correct dose.

Page 45

1. a) i) Phagocytes engulf the pathogen, then digest it. **(1 mark)**
      ii) Antibodies lock onto antigen/pathogen/clump pathogens together. **(1 mark)**
   b) Vaccine contains dead/heat-treated pathogen/microbe. **(1 mark)** Antigen recognised as foreign **(1 mark)**; lymphocytes produce antibodies against it **(1 mark)**; memory cells remain in system ready to produce antibodies if re-infection occurs **(1 mark)**.
   c) Antibiotics don't work against viruses **(1 mark)**; over-prescription may lead to antibiotic resistance **(1 mark)**.
   d) **Any one from:** antiviral drug/analgesic **(1 mark)**

2. a) Yes (no marks). Any two for **2 marks**: DDD results in a greater weight loss; 5.8 compared with 3.2; significantly higher than the placebo.
      No (no marks). Any two for **2 marks**: Number of volunteers for the DDD trial is very small compared with the other two trials; more trials need to be carried out; data is unreliable.
   b) In a double blind trial, neither the volunteers nor the doctors know which drug has been given **(1 mark)**; this eliminates all bias from the test/scientists cannot influence the volunteers' response in any way **(1 mark)**.

Page 47

1. light, touch/pressure, sound, chemicals (including taste and smell)
2. Maintenance of a constant internal environment.

Page 49

1. Axons/dendrites.
2. They ensure a rapid response to a threatening/harmful stimulus, e.g. picking up a hot plate. As a result, they reduce harm to the human body.
3. Flow diagram with the following labels: (stretch) receptor at knee joint stimulated; sensory neurone sends impulse to spine; intermediate/relay neurone relays impulse to motor neurone; motor neurone sends impulse to (thigh) muscle; (thigh) muscle contracts.

Page 51

1. It releases a range of hormones that control other processes in the body.
2. In the ovaries and the pituitary (female); in the testes (male).
3. Reduced ability of cells to absorb insulin and therefore high levels of blood glucose. This leads to tiredness, frequent urination, poor circulation, eye problems, etc.

Page 53

1. FSH acts on the ovaries, causing an egg to mature; oestrogen inhibits further production of FSH, stimulates the release of LH and promotes repair of the uterus wall; LH stimulates release of an egg; progesterone maintains the lining of the uterus after ovulation has occurred and inhibits FSH and LH.
2. Just after menstruation has stopped/from day 7.
3. **Any one from:** production of sperm in testes; development of muscles and penis; deepening of the voice; growth of pubic, facial and body hair.

Page 55

1. Oral contraceptive, hormone injection or implant.
2. They provide a barrier/prevent transferral of the virus during sexual intercourse.
3. **Any two from:** light or no menstruation; lasts many years; doesn't interrupt sexual activity.

Page 57

1. a) Stimulus. **(1 mark)**
   b) The response is automatic/unconscious **(1 mark)**; response is rapid **(1 mark)**.
   c) Stimulus at receptor triggers sensory neurone to send impulse **(1 mark)**; impulse received in brain and/or spinal cord and signal sent to intermediate/relay neurone **(1 mark)**; impulse then sent to motor neurone **(1 mark)**; motor neurone sends impulse to effector/muscles in hand **(1 mark)**.
2. a) The person's blood sugar level **(1 mark)** fluctuates dramatically **(1 mark)**.
   b) The person has eaten their breakfast at A and their lunch at B. **(1 mark)**
   c) There would be a slight rise in blood sugar level, followed by a swift drop back to normal level. **(1 mark)**
   d) Their blood sugar level had dropped too low/below normal. **(1 mark)**

Page 59

1. It allows variation, which gives an evolutionary advantage when the environment changes.
2. **Any one from:** produces clones of the parent – if these are successfully adapted individuals, then rapid colonisation and survival can be achieved; only one parent required; fewer resources required than sexual reproduction; faster than sexual reproduction.

Page 61

1. It has allowed the production of linkage maps that can be used for tracking inherited traits from generation to generation. This has led to targeted treatments for these conditions.
2. Proteins.

Page 63

1. A dominant allele controls the development of a characteristic even if it is present on only one chromosome in a pair. A recessive allele controls the development of a characteristic only if a dominant allele is not present.
2. A gene is a sequence of DNA code that determines the trait of an individual. Alleles are alternative forms of the same gene.

Page 65

1. Zero/0%.

Page 67

1. XX.
2. Causes production of thick mucus in respiratory pathways and interferes with enzyme production.

Page 69

1. **Any two from:** the fossil record; comparative anatomy; looking at changes in species during modern times; studying embryos and their similarities; comparing genomes of different organisms.
2. The conditions for their formation are rare, e.g. rapid burial and a lower chance of being discovered before they are eroded.

Page 71

1. **Evolution:** the long-term changes seen in species over a long period of time. **Natural selection:** the mechanism by which evolution occurs.
2. **Any two examples**, e.g. Kettlewell's moths; antibiotic resistance in bacteria.
3. They must be passed on by the well-adapted individual that contains them, through reproduction, to the next generation.

Page 72

1. **Similarity:** Beneficial /successful traits are passed on to the next generation; unsuccessful traits are removed from the species' population.
   Difference: Humans select the desired traits in selective breeding while the (natural) environment is the selective pressure in Natural Selection.

Page 73

1. Genetic engineering is more precise and it takes less time to see results.
2. **Reason for:** food production improved through increased yields and better nutritional content. **Reason against:** GM plants may spread their genetic material into the wider ecosystem, resulting in, for example, herbicide-resistant weeds/ possible harmful effect of GM foods on consumers.

Page 75

1. Genus and species.
2. Invertebrates, vertebrates, protists, higher plants.
3. Common ancestor.

Page 77

1. **Any four from:**
   - adapted to environment/had different characteristics
   - named examples of different characteristics, e.g. some horse-like mammals had more flipper-like limbs (as whales have flippers); idea of competition for limited resources
   - examples of different types of competition; idea of survival of the fittest
   - adaptations being advantageous to living in water/idea that adaptations helped them survive
   - named examples of different adaptations, e.g. some horse-like mammals had more flipper-like limbs that allowed them to swim well in water; idea of inheritance of successful characteristics
   - (named) characteristics/adaptations passed on (through breeding).
   (4 marks)
2. a) Ff
   b) ff
3. a) Nitrogenous base. **(1 mark)**
   b) Double helix. **(1 mark)**

Page 79

1. A **habitat** is the part of the physical environment in which an animal lives. An **ecosystem** includes the habitat, its communities of animals and plants, together with the physical factors that influence them.
2. **A mixed-leaf woodland:** the biodiversity is greater and so the links between different organisms are more extensive.
3. A **population** is the number of individuals of a species in a defined area. A **community** contains many different populations of species.
4. **Any two examples**, e.g. chemotrophs (live in deep ocean trenches and volcanic vents) and icefish (exist in waters less than 0°C in temperature).
5. So that they survive to reproductive age and pass on their genes to the next generation.

Page 81

1. Where you have a transition from one habitat to the next. It would tell you how numbers of different species vary along the line of the transect.

Page 83

1. The level that an organism feeds at, e.g. producer, primary consumer level. It refers to layers in a pyramid of biomass.
2. As the rabbit population decreases, there is less food available for foxes; so the fox population decreases. Fewer rabbits are therefore eaten, so the rabbit population then increases.

Page 85

1. **One from:** sulfur dioxide; nitrogen dioxide.
2. Numbers increase at a rapidly increasing rate.

Page 87

1. With more species, there are more relationships between different organisms and therefore more resistance to disruption from outside influences.
2. Plant a variety of trees, especially deciduous; protect rare species; make it a Site of Special Scientific Interest (SSSI); protect against invasive species, e.g. red squirrel.

Page 89

1. Condensation and evaporation.
2. As coal.
3. **Any two from:** combustion, animal respiration, plant respiration, microbial respiration (including decay).

Page 91

1. a) Respiration. **(1 mark)**
   b) **Any two from:** Fossil fuels represent a carbon 'sink'/they absorbed great quantities of carbon many millions of years ago from the atmosphere; combustion in power stations returns this carbon dioxide; less burning of fossil fuels cuts down on carbon emissions; alternative sources of energy may not return as much carbon dioxide to the atmosphere. **(2 marks)**
2. a) 120–125 **(1 mark)**
   b) When kingfishers increase, fish decrease/when kingfishers decrease, fish increase. **(1 mark)**
   c) **Any two from:** protection of rare species; captive breeding programs; maintaining habitat conditions, e.g. flooded wetland; creating artificially made habitats. **(2 marks)**
3. features; characteristics (either way around); suited; environment; evolutionary; survival. (**6 words correct = 3 marks, four or five words correct = 2 marks, two or three words correct = 1 mark, one or 0 words correct = 0 marks.**)

Chemistry

Page 93
1. A compound.
2. Simple distillation.

Page 95
1. The plum pudding model suggests that the electrons are embedded within the positive charge in an atom. The nuclear model suggested that the positive charge was confined in a small volume (the nucleus) of an atom.
2. Because most of the positive charge passed straight through the atom. As only a few alpha particles were deflected, this suggested that the positive part of the atom was very small.
3. 6 protons, 6 electrons and 7 neutrons.
4. Isotopes.

Page 97
1. 2, 8, 1
2. He predicted that there were more elements to be discovered. He predicted their properties and left appropriate places in the periodic table based on his predictions.
3. A metal.

Page 99
1. Because they have full outer shells of electrons.
2. Lithium hydroxide and hydrogen.
3. The reactivity decreases.
4. Potassium bromide and iodine.

Page 101
1. a) protons = 17 **(1 mark)**, electrons = 17 **(1 mark)**, neutrons = 18 **(1 mark)**.
   b) 2, 8, 7 **(1 mark)**.
   c) Similarity: same number of protons in each atom, same number of electrons in each atom, same chemical properties **(1 mark)**. Difference: different number of neutrons in each atom **(1 mark)**.
   d) **One from**: does not conduct heat/electricity; gas at room temperature **(1 mark)**.
   e) $Cl_2 + 2NaI \rightarrow 2NaCl + I_2$ (**1 mark for correct formula, 1 mark for correct balancing**).
   f) Chlorine is less reactive than fluorine **(1 mark)**, so it is unable to displace fluorine from a compound **(1 mark)**.
2. a) Filtration **(1 mark)**.
   b) Flask connected to condenser with bung/thermometer in place **(1 mark)**.
   Water flowing in at the bottom and leaving at the top **(1 mark)**.
   Salt water and pure water labelled **(1 mark)**.

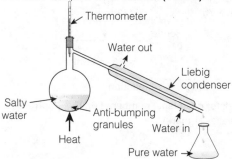

Thermometer
Water out
Liebig condenser
Salty water
Anti-bumping granules
Water in
Heat
Pure water

   c) The water is lost/evaporates during crystallisation **(1 mark)**.
3. $\frac{(10 \times 20) + (11 \times 80)}{100}$ **(1 mark)**
   = 10.8 **(1 mark)**
   Answer to 3 significant figures. **(1 mark)**

Page 103
1. Ionic.
2. Two.
3. A lattice (regular arrangement) of cations surrounded by a sea of (delocalised) electrons.

Page 105
1. Electrostatic forces/strong forces between cations and anions
2. Simple molecular, polymers, giant covalent
3.

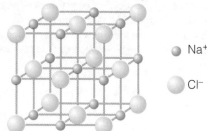

Na^+

Cl^-

Page 107
1. The strength of the forces acting between the particles present.
2. The electrostatic forces of attraction between the ions are strong.
3. The intermolecular forces/forces between the molecules.
4. Polymers have larger molecules therefore there are more forces between the molecules.
5. There are lots of strong covalent bonds that need lots of energy to break them all.

Page 109
1. Because there are strong forces of attraction between the metal cations and delocalised electrons.
2. In alloys, the layers of metal ions are not able to slide over each other.
3. Each carbon atom has a spare electron that allows it to conduct electricity.
4. **Two from**: high tensile strength; high electrical conductivity; high thermal conductivity.
5. **Two from**: drug delivery into the body; lubricants; reinforcing materials.

Page 111
1. a) Ionic **(1 mark)**
   b)

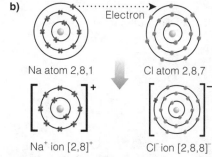

Electron
Na atom 2,8,1   Cl atom 2,8,7
Na^+ ion $[2,8]^+$ Cl^- ion $[2,8,8]^-$

   (**Correct electronic structure of atoms, 1 mark; correct electronic structure of ions, 1 mark; correct charges on ions, 1 mark.**)
   c) There are strong electrostatic forces of attraction **(1 mark)** between the ions **(1 mark)**.
   d) High **(1 mark)**. Lots of energy is needed to overcome the forces holding the ions together **(1 mark)**.
   e)

 — Negatively charged chloride ions

+ Positively charged sodium ions **(1 mark)**

2. **a)** Covalent **(1 mark)**.
   b) **i)** Two **(1 mark)**.
   ii) Simple molecular **(1 mark)**.
   c) **i)** Giant covalent (macromolecular) **(1 mark)**.
   ii) Simple molecular structures consist of many individual molecules held together by weak intermolecular forces **(1 mark)**. Giant covalent structures are lots of atoms all held together by covalent bonds **(1 mark)**.
   d) Silicon dioxide will have a higher boiling point **(1 mark)**. The covalent bonds in silicon dioxide that need to be broken in order to boil it are much stronger than the intermolecular forces that need to be broken in order to boil carbon dioxide **(1 mark)**.

Page 113

1. The same.
2. **a)** 80
   b) 58
3. Lower.

Page 115

1. $0.05 cm^3$
2. Average = 21.50 cm^3; Range = 1.80 cm^3

Page 117

1. **a)** Lower **(1 mark)**, because the $CaCO_3$/calcium carbonate will have thermally decomposed meaning that some of the $CaCO_3$ will have decomposed into carbon dioxide, which will have gone into the air **(1 mark)**.
   b) 100 **(1 mark)**.
   c) 100 g of $CaCO_3$ forms 44 g of CO_2; therefore 10 g of $CaCO_3$ will have formed 4.4 g of CO_2 **(1 mark for working, 1 mark for the correct answer based on working; correct answer on its own scores 2 marks. The final answer must include the units)**.
2. **a)** Oxygen from the air reacted with the copper **(1 mark)**.
   b) 0.05g **(1 mark)**
   c) copper + oxygen → copper (II) oxide. (Allow copper oxide) **(1 mark for reactants, 1 mark for product)**.
   d) The mass of the copper increased and the mass of the air decreased by the same amount.
3. **a)** (21 ÷ 75) x 100 = 28g **(1 mark)**
   b) (21 ÷ 75) x 1000 = 280 g/dm^3 **(1 mark)**

Page 119

1. magnesium + oxygen → magnesium oxide
2. Sodium (because it is more reactive).
3. **Two from**: zinc, iron, copper.

Page 121

1. zinc + sulfuric acid → zinc sulfate + hydrogen.
2. A metal, metal oxide or metal carbonate.
3. Lithium nitrate.
4. Add solid until no more reacts, filter off the excess solid, crystallise the remaining solution.

Page 123

1. An acid.
2. OH^-
3. $H^+_{(aq)} + OH^-_{(aq)} \rightarrow H_2O_{(l)}$
4. So that the ions are able to move.

Page 125

1. Potassium will be formed at the cathode; iodine will be formed at the anode.
2. Hydrogen will be formed at the cathode; iodine will be formed at the anode.

Page 127

1. It cools down.
2.

3. Endothermic

Page 129

1. **a)** $2 Zn_{(s)} + O_{2(g)} \rightarrow 2 ZnO_{(s)}$ (state symbols are not required. **1 mark awarded for the correct symbols/formulae and for the equation being correctly balanced**).
   b) ZnO/zinc oxide **(1 mark)**, because it loses oxygen **(1 mark)**.
   c) Less reactive **(1 mark)**; magnesium is able to displace zinc from zinc oxide, meaning that magnesium is more reactive than zinc **(1 mark)**.
   d) By reduction with carbon **(1 mark)**, because it is below carbon in the reactivity series **(1 mark)**.
2. **a)**

   (1 mark for each label numbered on the graph.)
   b) It increases **(1 mark)**

Page 131

1. 0.75 g/s
2. **Two from**: the concentrations of the reactants in solution; the pressure of reacting gases; the surface area of any solid reactants; temperature; presence of a catalyst.
3. By attaching a gas syringe and recording the volume of gas collected in a certain amount of time, e.g. volume collected every 10 seconds.
4. The rate increases.
5. The idea that, for a chemical reaction to occur, the reacting particles must collide with sufficient energy.
6. The minimum amount of energy that the particles must have when they collide in order to react.
7. There are more particles in the same volume of liquid and so there are more chances of reactant particles colliding.

Page 133

1. A species that speeds up a chemical reaction but is not used up during the reaction.
2. They provide an alternative pathway of lower activation energy.
3. $\rightleftharpoons$
4. Endothermic.

Page 135

1. a) 50 seconds **(1 mark)**, as after this time no more gas was collected/the volume of gas did not change **(1 mark)**.
   b) $40 \div 50 = 0.8$ cm^3/s (working **1 mark**: allow $40 \div$ answer to part a; answer with units, **1 mark**).
   c) The gradient was steeper after 10 seconds than after 40 seconds. **(1 mark)**
   d)
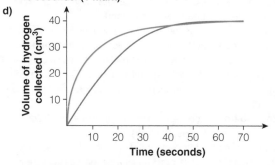

   (**Curve is steeper than in original graph, 1 mark**; **final volume is 40 cm^3, 1 mark**.)
   e) At a higher concentration there will be more particles of acid in the same volume of solution **(1 mark)**, and so there will be more/an increased probability/more likelihood of/collisions **(1 mark)**.
2. a) The water in the equation appears as a gas **(1 mark)**.
   b) That the reaction is reversible **(1 mark)**.

Page 137

1. Dead biomass.
2. Molecules that contain carbon and hydrogen atoms only.
3. C_3H_8
4. Fractional distillation.

Page 139

1. $C_2H_6 + 3.5\ O_2 \rightarrow 2CO_2 + 3H_2O$ **or** $2C_2H_6 + 7\ O_2 \rightarrow 4CO_2 + 6H_2O$
2. C_4H_{10}
3. By heating with steam
4. Add bromine water. Alkenes decolourise bromine water / turn it from orange to colourless.
5. To make plastics / polymerisation

Page 141

1. a) Dead biomass sinks to ocean bottom **(1 mark)**; decaying biomass is covered in mud, which turns to rock **(1 mark)**; biomass decays, slowly forming crude oil under the rock **(1 mark)**.
   b) Fractional distillation **(1 mark)**; the molecules in crude oil are separated according to their boiling points **(1 mark)**.
   c) Saturated **(1 mark)**; hydrocarbons/molecules with the general formula C_nH_{2n+2} **(1 mark)**.
   d) Cracking turns relatively useless long-chain molecules **(1 mark)** into more useful products **(1 mark)**.
   e) By passing the hydrocarbon vapour over a hot catalyst/mixing the hydrocarbon with steam at high temperatures **(1 mark)**.
   f) C_4H_{10} **(1 mark)**.
2. a) Kerosene – aircraft fuel / fuel for stoves **(1 mark)**; Bitumen – tar for roads / roofing **(1 mark)**
   b) bitumen **(1 mark)**
   c) kerosene **(1 mark)**
   d) bitumen **(1 mark)** because the molecules in bitumen are larger / stronger intermolecular forces **(1 mark)**
3. a) Molecules that contain carbon and hydrogen **(1 mark)** only **(1 mark)**.
   b) **Any two from:** differ by CH_2 in their molecular formula from neighbouring compounds; show a gradual trend in physical properties; have similar chemical properties. **(2 marks)**
   c) 17 **(1 mark)**
   d)

$$H-\overset{\displaystyle \overset{H}{|}}{\underset{\displaystyle \underset{H}{|}}{C}}-\overset{\displaystyle \overset{H}{|}}{\underset{\displaystyle \underset{H}{|}}{C}}-\overset{\displaystyle \overset{H}{|}}{\underset{\displaystyle \underset{H}{|}}{C}}-H \quad \text{(1 mark)}$$

   e) $C_4H_{10\,(g)} + 6\frac{1}{2}\ O_{2\,(g)} \rightarrow 4CO_{2\,(g)} + 5H_2O_{\,(l)}$ **(Ignore state symbols)** (**Allow any correct multiple**, e.g.: $2C_4H_{10\,(g)} + 13O_{2\,(g)} \rightarrow 8CO_{2\,(g)} + 10H_2O_{\,(l)}$) (**1 mark for correct formula of reactants and products; 1 mark for correct balancing**)
   f) Add bromine water / bubble the gases through bromine water **(1 mark)**.
   The alkene will decolourise the bromine water / turn it from brown / orange to colourless. **(1 mark)**
   The bromine water remains orange in the alkane **(1 mark)**

Page 143

1. A single element or compound.
2. A mixture that has been designed as a useful product.
3. **Two from**: fuels; cleaning materials; paints; medicines; foods; fertilisers.
4. 0.55

Page 145

1. Oxygen.
2. Add a lit splint and there will be a squeaky pop.
3. Calcium hydroxide solution.
4. It turns milky.
5. It turns it red before bleaching it.

Page 147

1. a) Z **(1 mark)**. There are three 'spots' on the chromatogram of Z **(1 mark)**.
   b) W and Z **(1 mark)**
   c) Red and yellow **(1 mark)**
   d) X **(1 mark)**. None of the spots in X are at the same height of the spots in the red, yellow or blue ink **(1 mark)**.
   e) $\frac{22.5}{29}$ **(1 mark)** = 0.78 (allow between 0.76 and 0.79) **(1 mark)**
   f) W **(1 mark)**. It contains a single compound **(1 mark)**.
   g) **Any one from:** a mixture that has been designed as a useful product; a mixture made by mixing the individual components in carefully measured quantities. **(1 mark)**

2.

Gas	Test	Observation
Hydrogen	Add a lit splint	A 'squeaky pop' is heard
Oxygen	Add a glowing splint	Splint relights
Chlorine	Add moist blue litmus paper	Litmus paper turns red and is then bleached / turns white
Carbon dioxide	Add limewater	The limewater turns milky / cloudy

(1 mark for each correctly filled in box up to a maximum of 8 marks)

Page 149

1. **Two from**: carbon dioxide; water vapour; methane; ammonia; nitrogen.
2. As a product of photosynthesis.
3. Approximately $\frac{4}{5}$ or 80%.
4. Because it was used in photosynthesis and to form sedimentary rocks.

Page 151

1. **Two from**: water vapour; carbon dioxide; methane.
2. Increased animal farming/rubbish in landfill sites.
3. **Two from**: rising sea levels leading to flooding/coastal erosion; more frequent/severe storms; changes to the amount, timing and distribution of rainfall; temperature and water stress for humans and wildlife; changes in the food producing capacity of some regions; changes to the distribution of wildlife species.
4. **One from**: disagreement over the causes and consequences of climate change; lack of public information and education; lifestyle changes, e.g. greater use of cars and aeroplanes; economic considerations, i.e. the financial costs of reducing the carbon footprint; incomplete international cooperation.

Page 153

1. **Two from**: carbon dioxide; carbon monoxide; water vapour; sulfur dioxide; nitrogen oxides.
2. From the incomplete combustion of fossil fuels.
3. They cause respiratory problems and can form acid rain.

Page 155

1. Living such that the needs of the current generation are met without compromising the ability of future generations to meet their own needs.
2. Pure water has no chemicals added to it. Potable water may have other substances in it but it is safe to drink.
3. Filtered and then sterilised.
4. Distillation or reverse osmosis.

Page 157

1. The environmental impact of a product over the whole of its life.
2. **Two from**: how much energy is needed; how much water is used; what resources are required; how much waste is produced; how much pollution is produced.
3. Some of the values are difficult to quantify, meaning that value judgements have to be made which could be biased/based on opinion.
4. **One from**: using fewer items that come from the earth; reusing items; recycling more of what we use.

Page 159

1. a) Volcanoes **(1 mark)**.
   b) Nitrogen **(1 mark)**.
   c) Oxygen is formed by photosynthesis **(1 mark)**; oxygen has not always been present in the atmosphere because green plants/algae have not always existed **(1 mark)**.
   d) Carbon dioxide is absorbed into the oceans/forms carbonate rocks **(1 mark)**, and is used in photosynthesis **(1 mark)**.
   e) There has been increased consumption of fossil fuels **(1 mark)**, e.g. since the Industrial Revolution/greater use of transport such as cars which produce carbon dioxide when they burn fuel **(1 mark)**.

2. a) **Any two from:** increased animal farming; deforestation; rubbish in landfill sites. **(2 mark)**
   b) A measure of the total amount of carbon dioxide (and other greenhouse gases) **(1 mark)** emitted over the life cycle of a product, service or event **(1 mark)**
   c) **Any two from:** use of alternative energy supplies; energy conservation; CCS (carbon capture and storage); carbon taxes or licences; carbon offsetting measures; carbon neutrality. **(2 mark)**
   d) i) **Any one from:** carbon monoxide; oxides of nitrogen; sulfur dioxide. **(1 mark)**
      ii) **Either from:** carbon monoxide – toxic gas that reduces the ability of blood to carry oxygen; sulfur dioxide/oxides of nitrogen – cause respiratory problems in humans and animals / can form acid rain. **(1 mark)**

Physics

1. A scalar quantity only has a magnitude, a vector quantity has both a magnitude and a direction.
2. **Two from**: friction; air resistance; tension; normal contact force
3. $0.067 \times 10 = 0.67$ N
4. 78 J

1. $0.1 \times 2 = 0.2$ N
2. Any point where the limit of proportionality hasn't been exceeded
3. Elastic potential energy

1. Distance has a magnitude but not a direction.
2. $\frac{10}{2.5} = 4$ km/h
3. Speed is a scalar quantity as it only has a magnitude. Velocity is speed in a given direction. Velocity has a magnitude and a direction so it is a vector quantity.

1. Speed
2. A negative acceleration

1. 67 N
2. $89 \times 10 = 890$ N
3. It continues to move at the same speed.

1. Thinking distance and braking distance
2. **Two from**: rain; ice; snow

1. a) weight = mass × gravitational field strength
   $3.7 \times 187 = 691.9$ N **(2 marks)**
   b) Yes. The rover would weigh more (1870 N) **(1 mark)** as the gravitational field strength on Earth is greater than that on Mars **(1 mark)**
2. a) i) Gradient of graph from 0–3 s = $\frac{25}{3} = 8.33$
   Acceleration = 8.33 m/s^2 **(2 marks)**
   ii) Area under graph 0–3 s = $0.5 \times (3 \times 25) = 37.5$
   Distance travelled = 37.5 m **(2 marks)**
   b) Yes **(1 mark)** as the line on the graph is steeper from 0–3 than 3–6 / A steeper line indicates a faster acceleration **(1 mark)**
3. a) The upward force from the ground is 890 N **(1 mark)** as it is equal and opposite to the force the cyclist is exerting on the ground **(1 mark)**.
   b) As the forces on the rider are balanced **(1 mark)** there is no resultant force and the rider would continue to move at a constant speed. This is Newton's first law **(1 mark)**.
   c) There would be a resultant force backwards **(1 mark)** so the rider would slow down **(1 mark)**.
4. resultant force = mass × acceleration
   $= 1200 \times 2.5 = 3000$ N or 3 kN **(2 marks)**

1. $0.5 \times 0.065 \times 6^2 = 11.7$ J
2. $17 \times 987 \times 10 = 167.79$ kJ
3. It is the amount of energy required to raise the temperature of one kilogram of a substance by one degree Celsius.

1. To ensure more energy is usefully transferred and less is wasted
2. $\frac{400}{652} = 0.61 = 61\%$
3. It reduces the thermal conductivity.

1. Renewable. **Three from**: bio-fuel; wind; hydro-electricity; geothermal; tidal power; solar power; water waves
   Non-renewable. **Three from**: coal; oil; gas; nuclear fuel
2. The wind doesn't always blow and it's not always sunny, so electricity isn't always generated.
3. They produce carbon dioxide, which is a greenhouse gas. Increased greenhouse gas emissions are leading to climate change. Particulates and other pollutants are also released, which cause respiratory problems.

1. a) Renewable resources can be replenished as they are used **(1 mark)** whilst non-renewable energy resources will eventually run out **(1 mark)**
   b) Accidents at nuclear power stations could have devastating health and environmental effects **(1 mark)**. Nuclear waste is very hazardous **(1 mark)**.
   c) **Two from**: building tidal power stations involves the destruction of habitats; wind energy doesn't produce electricity all the time; some people think wind turbines are ugly and spoil the landscape **(2 marks)**
2. a) kinetic energy = 0.5 × mass × (speed)^2
   $= 0.5 \times 1600 \times 32^2$
   $= 819\ 200$ J or 819.2 kJ **(2 marks)**
   b) efficiency = $\frac{\text{useful output energy transfer}}{\text{useful input energy transfer}}$
   $= \frac{819.2}{1500} = 0.55$ or 55% **(2 marks)**
   c) Sound **(1 mark)** and heat **(1 mark)**
   d) **One from**: lubrication; oil **(1 mark)**
3. a) g.p.e. = mass × gravitational field strength × height
   $= 77 \times 150 \times 10$
   $= 115.5$ kN **(2 marks)**
   b) kinetic energy = 0.5 × mass × (speed)^2
   $= 0.5 \times 77 \times 20^2$
   $= 15.4$ kN **(2 marks)**
   c) Gravitational potential energy to kinetic energy **(1 mark)** to elastic potential energy **(1 mark)**

1. Longitudinal: Sound wave
   Transverse: **One from**: water wave; electromagnetic wave
2. The distance from a point on one wave to the equivalent point on an adjacent wave
3. 1792 m/s

1. Infrared radiation
2. Visible light
3. Mutation of genes and cancer
4. **One from**: energy efficient lamps; sun tanning

1. a) Compression is a region in a longitudinal wave where the particles are closer together **(1 mark)**. Rarefaction is a region in a longitudinal wave where the particles are further apart **(1 mark)**

b) frequency = $\dfrac{\text{wave speed}}{\text{wavelength}}$

$$= \dfrac{200}{3.2}$$

$$= 62.5 \text{ Hz } \textbf{(2 marks)}$$

2. a) Ultraviolet radiation **(1 mark)**

b) It can increase the chance of developing skin cancer. **(1 mark)**

c) Human eyes can only perceive a limited range of electromagnetic radiation (visible light) **(1 mark)**

3. Sound travels as a **longitudinal** wave. When a sound wave travels in air the oscillations of the air particles are **parallel** to the direction of energy transfer. The sound wave **doesn't transfer matter, only energy**. **(3 marks)**

Page 189

1. The symbol for a variable resistor has an arrow through it.

2. 4.2 C

3. 90 A

Page 191

1. Because resistance depends on light intensity

2. An ammeter and a voltmeter

3. 8 V

4. 2 Ω

Page 193

1. 5 A

2. Add the resistances of both resistors together

3. Lower

Page 195

1. Blue

2. The earth wire is a safety wire to stop the appliance becoming live.

3. Touching a live wire would produce a large potential difference across the body.

Page 197

1. $5^2 \times 2 = 50$ W

2. $60 \times 12 = 720$ J

3. They lower the potential difference of the transmission cables to a safe level for domestic use.

Page 199

1. They would repel

2. When it's placed in a magnetic field

3. At the poles of the magnet

4. The Earth's core is magnetic and produces a magnetic field.

5. The current through the wire and the distance from the wire

6. The same shape as the magnetic field around a bar magnet

Page 201

1. a) current = $\dfrac{\text{potential difference}}{\text{resistance}}$

$$= \dfrac{15}{3}$$

$$= 5 \text{ A } \textbf{(3 marks)}$$

b) 5 A **(1 mark)**. The current is the same at all points in a series circuit **(1 mark)**.

c)

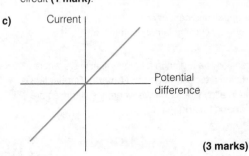

(3 marks)

d) **One from**: filament lamp; LDR; diode; thermistor **(1 mark)**

2. a) Brown **(1 mark)**

b) **Any two points from:** The live wire carries the alternating potential difference from the supply / whilst the neutral wire completes the circuit and is at earth potential. / Mixing the two up could lead to damage to the appliance or injury. **(2 marks)**

3. a)

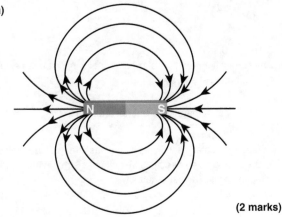

(2 marks)

b) The magnets would attract **(1 mark)**.

c) An induced magnet is only magnetic when placed in a magnetic field whilst a bar magnet is always magnetic **or** An induced magnet always causes a force of attraction while a bar magnet can cause a force of attraction or repulsion. **(2 marks)**

4. a) If an electrical current passes through the wire. **(1 mark)**

b) It would increase the strength of the magnetic field produced by the wire. **(1 mark)**

c) If the wire was wrapped around an iron core. **(1 mark)**

Page 203

1. When the molecules collide with the wall of their container they exert a force on the wall, causing pressure.

2. $\dfrac{2}{0.002} = 1000$ kg/m³

3. The mass stays the same

Page 205

1. $200 \times 126 = 25\ 200$ kJ

2. Specific latent heat of fusion is the energy required for a change of state from solid to liquid. Specific latent heat of vapourisation is the energy required for a change of state from liquid to vapour.

3. It is changing state.

Page 207

1. a) density = $\dfrac{\text{mass}}{\text{volume}}$

$$= \dfrac{0.15}{0.0001}$$

$$= 1500 \text{ kg/m}^3 \textbf{ (2 marks)}$$

b) The mass remains constant **(1 mark)** and the density decreases **(1 mark)**.

c) The temperature would remain constant as it was changing state **(1 mark)** and the solid's internal energy would increase **(1 mark)**.

d) energy for a change of state = mass × specific latent heat

$$= 0.15 \times 574\ 000$$

$$= 86.1 \text{ kJ or } 86100 \text{ J } \textbf{(2 marks)}$$

2. a) The particles gain kinetic energy as the temperature increases **(1 mark)**. This increases the speed at which the particles collide with the sides of the container they are in **(1 mark)**.

b) The same number of particles occupy a greater volume **(1 mark)**, reducing the rate of collisions with the sides of the container **(1 mark)**.

3. Specific latent heat of vaporisation **(1 mark)**

Page 209

1. The electrons become excited and move to a higher energy level, further from the nucleus.
2. Isotopes
3. The plum pudding model suggested the atom was a ball of positive charge with negative electrons embedded in it. The nuclear model has a nucleus with electrons orbiting.

Page 211

1. A few centimetres
2. **Accept any dense material**, e.g.: concrete or lead.
3. The mass of the nucleus doesn't change but the charge of the nucleus does change.

Page 213

1. When the time taken for the number of nuclei in a sample of the isotope halves, or when the count rate (or activity) from a sample containing the isotope falls to half of its initial level.
2. 5 months
3. The radioactive atoms decay and release radiation.

Page 215

1. a) i) 22 **(1 mark)**
      ii) 22 **(1 mark)**
      iii) 26 **(1 mark)**
   b) i) Number of protons is the atomic number **(1 mark)**, which is the bottom number on the symbol **(1 mark)**.
      ii) The number of electrons equals the number of protons **(1 mark)**. The number of protons is 28 **(1 mark)**.
      iii) The number of neutrons is the mass number **(1 mark)**. The number of protons is 48 − 22 = 26 **(1 mark)**.
2. a) The plum pudding model suggested that the atom is a ball of positive charge with negative electrons embedded in it **(1 mark)**. Rutherford, Geiger and Marsden said the mass of an atom was concentrated in the nucleus with electrons orbiting the nucleus **(1 mark)**.
   b) Niels Bohr suggested that the electrons orbit the nucleus at specific distances **(1 mark)**. Chadwick provided the evidence for the existence of the neutron within the nucleus **(1 mark)**.
3. a) 60 days **(1 mark)**. This is the point where the counts per second have halved (Also accept $\frac{800}{2}$ = 400) **(1 mark)**
   b) No **(1 mark)** because its half-life is too long so it would persist in the body for too long **(1 mark)**.

Notes

Notes

Notes

Notes

Index

Index

The Periodic Table

Key

- ▉ Metals
- ▉ Non-metals

Key to element box:

Relative atomic mass → 1
Atomic symbol → **H**
Name → hydrogen
Atomic number → 1

1	2											3	4	5	6	7	0 or 8
																	4 **He** helium 2
7 **Li** lithium 3	9 **Be** beryllium 4											11 **B** boron 5	12 **C** carbon 6	14 **N** nitrogen 7	16 **O** oxygen 8	19 **F** fluorine 9	20 **Ne** neon 10
23 **Na** sodium 11	24 **Mg** magnesium 12											27 **Al** aluminium 13	28 **Si** silicon 14	31 **P** phosphorus 15	32 **S** sulfur 16	35.5 **Cl** chlorine 17	40 **Ar** argon 18
39 **K** potassium 19	40 **Ca** calcium 20	45 **Sc** scandium 21	48 **Ti** titanium 22	51 **V** vanadium 23	52 **Cr** chromium 24	55 **Mn** manganese 25	56 **Fe** iron 26	59 **Co** cobalt 27	59 **Ni** nickel 28	63.5 **Cu** copper 29	65 **Zn** zinc 30	70 **Ga** gallium 31	73 **Ge** germanium 32	75 **As** arsenic 33	79 **Se** selenium 34	80 **Br** bromine 35	84 **Kr** krypton 36
85 **Rb** rubidium 37	88 **Sr** strontium 38	89 **Y** yttrium 39	91 **Zr** zirconium 40	93 **Nb** niobium 41	96 **Mo** molybdenum 42	[98] **Tc** technetium 43	101 **Ru** ruthenium 44	103 **Rh** rhodium 45	106 **Pd** palladium 46	108 **Ag** silver 47	112 **Cd** cadmium 48	115 **In** indium 49	119 **Sn** tin 50	122 **Sb** antimony 51	128 **Te** tellurium 52	127 **I** iodine 53	131 **Xe** xenon 54
133 **Cs** caesium 55	137 **Ba** barium 56	139 **La*** lanthanum 57	178 **Hf** hafnium 72	181 **Ta** tantalum 73	184 **W** tungsten 74	186 **Re** rhenium 75	190 **Os** osmium 76	192 **Ir** iridium 77	195 **Pt** platinum 78	197 **Au** gold 79	201 **Hg** mercury 80	204 **Tl** thallium 81	207 **Pb** lead 82	209 **Bi** bismuth 83	[209] **Po** polonium 84	[210] **At** astatine 85	[222] **Rn** radon 86
[223] **Fr** francium 87	[226] **Ra** radium 88	[227] **Ac*** actinium 89	[261] **Rf** rutherfordium 104	[262] **Db** dubnium 105	[266] **Sg** seaborgium 106	[264] **Bh** bohrium 107	[277] **Hs** hassium 108	[268] **Mt** meitnerium 109	[271] **Ds** darmstadtium 110	[272] **Rg** roentgenium 111							

Elements with atomic numbers 112–116 have been reported but not fully authenticated

*The lanthanoids (atomic numbers 58–71) and the actinoids (atomic numbers 90–103) have been omitted.

The relative atomic masses of copper and chlorine have not been rounded to the nearest whole number.